A SEASON IN THE SUN

A SEASON
IN THE SUN

THE INSIDE STORY OF BRUCE ARIANS, TOM BRADY,
AND THE MAKING OF A CHAMPION

LARS ANDERSON

WM
WILLIAM MORROW
An Imprint of HarperCollinsPublishers

HarperCollins books may be purchased for educational, business, or sales promotional use. For information, please email the Special Markets Department at SPsales@harpercollins.com.

FIRST EDITION

Designed by Elina Cohen

Library of Congress Cataloging-in-Publication Data has been applied for.

ISBN 978-0-06-316020-0

21 22 23 24 25   LSC   10 9 8 7 6 5 4 3 2 1

For my beautiful kids, Lincoln, Autumn, and Farrah.
Know that daddy loves you, always.

CONTENTS

||||||||||||||||||||||||||||||||||||||||||

CONTENTS

I don't think the beautiful reality of it truly hit me until we were cruising down the Hillsborough River during the victory boat parade. I was holding the Lombardi Trophy when, all of a sudden, "We Are the Champions" blasted from a speaker. With the song by Queen thumping on my chest, my eyes welled with tears. That's when I fully embraced it. We were the NFL champions of 2020. This book by Lars Anderson, who I have known and trusted for years, vividly tells the story of this very special season.

For me, the proudest part of the year was the collective job the players and the coaches did in beating COVID. The commitment they made, all those young guys, was incredible. All they did was go to work, go home, go to work, go home. We had only two players test positive for the coronavirus—that was unbelievable. Then we advance to the Super Bowl and safety Mike Edwards tests positive. Guess what? It was Mike Edwards Sr. They didn't check the birth certificate and it was Mike's dad. Nearly gave me a damned heart attack. I've never seen anything close to the challenge we all faced because of COVID. We had to meet differently, we had to eat differently, we had to practice differently. I kept telling the team, "Don't be the guy that screws up."

By about the middle of September, I told the players I was tired of being the head of the mask police, walking through the locker room and telling them to get your mask on. I emphasized that the guys either had to grow up and make a commitment to each other, or this wasn't gonna work.

Then the veteran guys took over. From that point on, I never had to say another word. The same thing happened after the loss in Chicago. I'd been bitching about penalties for twenty games. Why did it stop? Because they finally had enough. That game turned my stomach.

Of course, you can't say enough about Tom Brady and what he did for our entire franchise. What I learned is he is one hell of a teammate. You see the glitz and the glamour, you already know he's a leader, but I didn't know he would take a young guy aside and say, "You don't know how to take care of your body." When our great second-year linebacker Devin White didn't make the Pro Bowl, Devin was devastated. It was Tom who told Devin, "Hey, we've got bigger fish to fry." He got Devin out of the funk.

Tom reads his teammates so well. When we signed Tom as a free agent, linebacker Lavonte David approached me and said, "Coach, ain't no doubt we're gonna win." Lavonte is one of my main guys in the locker room. He said he could feel a difference in confidence the moment the team stepped onto the practice field for the first time with Tom. We just had to stay healthy.

I've got to give our owners so much credit. The Glazers took a chance on me when they hired me in 2019. It wasn't like I was this young up-and-coming thirty-something offensive savant—this seems to be what most owners are now looking for when hiring head coaches. At the time I was sixty-eight. But I had a proven track record—I'd twice been named the Associated Press Coach of the Year—and I had a plan. Put simply, they believed in me. I cannot express the gratitude I and my family have for the Glazer family.

My assistants—and trust me, there are at least five guys who will one day be head coaches on this staff—are family. They did a great job with all those Zoom calls during our virtual offseason and then during the season. I cannot teach over Zoom. I didn't do a single Zoom meeting the entire time, but our coaches were great in developing our young players.

I always go back to our staff because I wouldn't have taken this job if they weren't available. People don't realize how good this staff is. Byron Leftwich gets absolutely no credit as offensive coordinator because they give it to me and Tom. But Byron put our game plan together with the rest of the offensive coaches. I'll have my hands in it. The job he did with Tom was unbelievable.

Tom had learned a certain way for twenty years. Byron and I coached a totally different way. Tom was trying to understand our playbook and how we coached while we were busy blending our offense with the things he was most comfortable doing. In other words, we had a lot going on. But Byron did a wonderful job of working with Tom. Byron should be on the short list for any general manager looking for a head coach in 2022.

Todd Bowles, our defensive coordinator, has a hell of a track record. He was named the league's Assistant Coach of the Year when he was with me in Arizona. He'll be a head coach again, especially after his masterful game plan in the Super Bowl that shut down the Kansas City Chiefs. I've known Todd since he played for me at Temple, and he's the defensive version of me: aggressive, risk-taking, and absolutely fearless.

I knew developing our young players would be huge. In 2019, I told the Glazers I wanted to have a large coaching staff. I said we were going to have two practices going at the same time in the spring. I wanted the veterans on one field, the young players on another. So when they got to training camp, they were ready to go. The Glazers let me have the staff I wanted.

Mike Caldwell (inside linebackers coach) was with me in Arizona, Nick Rapone (safeties) coached for me at Temple. Kevin Ross (cornerbacks) and Todd both played for me at Temple. Kacy Rodgers (defensive line) was with Todd when he was the head coach of the New York Jets. Joe Gilbert (offensive line) was with me at Indy. Keith Armstrong (special teams coordinator) was a captain for me at Temple. Larry Foote (linebackers) played for me at Pittsburgh and was on my staff in Arizona. I hired Cody Grimm as a defensive/special teams assistant; he is going to be a hell of a coach. Even before they arrived in Tampa, all of those defensive coaches knew the playbook. This is a family. We are a family. Nobody's gonna stab me in the back because the rest of them would kick his ass.

We found a gem in Thaddeus Lewis, a former NFL quarterback who played at Duke. This kid's gonna be a star. He was a coaching intern with us last year and now he's going to be an assistant wide receivers coach. The NFL needs to do a better job of giving minority coaching candidates chances, and the internship program that I've promoted has given so many guys like Byron and Thaddeus that chance. This needs to be a league-wide policy.

We have a lot of diversity on our staff and that's no accident. It starts with the Glazers. They believe strongly in it, obviously, with Darcie Glazer Kassewitz being a co-owner with her brothers. On our staff we have age differences (one of our offensive consultants, Tom Moore, who has been with me forever, is eighty-two and is one of my most trusted confidants), racial differences, and gender differences. We're the only team with a Black offensive coordinator (Leftwich), a Black defensive coordinator (Bowles), a Black special teams coordinator (Armstrong), and a Black assistant head coach (Harold Godwin). All four of these guys should be head coaches now, and I can't emphasize enough that we wouldn't have won the Super Bowl without the contributions of each and every one of my staff members. Loyalty is huge to me—it's honestly more important than money—and these are some of the most loyal men I've had the privilege to be around. By my side through this special season was Mike Chiurco, the assistant to the head coach, who is another one of my most loyal dudes.

We are also the only team in the league with two full-time female staff members: Lori Locust, our assistant defensive line coach, and Maral Javadifar, our assistant strength and conditioning coach. Women teach in different ways and football's nothing but teaching. Back in early 2019 I went to speak at the University of Alabama–Birmingham (UAB) and bumped into Joe Pendry, who I once coached with at Kansas City. He asked me if I was looking for a female coach and said he knew a great one. He then put me in touch with Lori, who we brought in for an interview. She was a student when I coached at Temple back in the 1980s and she knows all my guys—we jokingly call our staff "Temple South." She fit in right away. The coaches love her. Ndamukong Suh loves her. And everyone feels the same way about Maral Javadifar. I swear she must have eight degrees, and as soon as the players realized she knew her stuff, it was like the issue of gender just went away.

I've got to give Jason Licht, our general manager, a ton of credit. First of all, they don't have me coaching in Tampa if Jason isn't here. I worked with him before in Arizona and he's a great evaluator of talent. He knows how to get coaches and scouts to work together. I thought we killed it in the draft that first year (2019), and in free agency we landed Shaq Barrett. That ain't bad. We didn't know how good Shaq would be, but we knew we had a slippery dude. Then we signed Suh, then we drafted Devin White in the first round. This doesn't happen without Jason. I trust him totally. He knew our staff and knew what our guys were looking for.

I've stepped back a bit from some of the multiple roles I used to take on as head coach—namely, those of offensive coordinator and play caller—and it's been fun. The negative is that I love calling plays. I trusted Byron and probably wouldn't have taken the job if he wasn't available. Now, I get to be around our defensive players more. I probably watched more college tape than I've ever watched and I talk to Jason all the time.

Stepping back also helped me be on top of the mood of our players. I don't allow pouting and I don't allow bitching. You want to light my fuse? Just show bad body language on the field. I quickly give you two choices: change or ask me to cut your ass.

|||||||||||||

Going into the draft in April 2020, we loved offensive tackle Tristan Wirfs out of Iowa and safety Antoine Winfield from Minnesota because they were high-character, mature guys. Tristan got a rude awakening from defensive ends Jason Pierre-Paul and Shaq Barrett in training camp. We had to tell them to let Tristan win now and then. In camp, he was overwhelmed, but they saw how good he could be and our defensive players started coaching him. Things like, "When you put that hand over there, I'm gonna do this to you." They helped his development as much as our coaches.

I knew Tristan was going to be special after the opener in New

Orleans. Right out of the box, he faced Cam Jordan and Trey Hendrickson, who's way better than people give him credit for.

Wirfs held his own with no help. Winfield made a play or two, but you saw in camp that he was gonna be a playmaker. After Game 1, we knew we had two rookies who were really special.

But there's no getting around it, signing Tom was the key to the whole thing.

I was confident that we had a great young team. We should have been in the playoffs the year before. We blew games. That 2019 season still pisses me off.

When I became the head coach at Arizona in 2013, we went 10–6 and almost made the playoffs in our first season. That 2019 Buc team was better by far. After the season, the Glazers asked me whether we could fix quarterback Jameis Winston. I said it's gonna be harder than I thought it would, because of the regression he experienced in December.

Talking to Tom during free agency, he knew how good we were defensively. He said he was really interested in Tampa, but he also had some other options. The Glazers were really excited, and that was important. They told us that if we thought Tom could still play, go for it. They knew what it would do for the city and the franchise. They knew if Tom came here, the Bucs would get talked about every day, and it would be because of one guy. Tom could have asked for more money and he would have gotten it.

I didn't need the validation of a Super Bowl to know that I'm a good NFL head coach, but I'd be lying if I said winning the Vince Lombardi Trophy didn't matter. My son, Jake, said it best: "You're now a Super Bowl champ, but you're not a smarter coach." That's true, but once you win it all, you can look back at some people who didn't believe in your ability, and say, "Screw you."

By the end of the season, we were a hell of a football team. When the Super Bowl confetti was coming down, I'll always remember being on the field when my kids and my wife came down and we shared a hug. That was the best moment, and we now have a big print of our family hug that hangs in our living room; I look at it every single day. I didn't even see the team after the game. When I came back through the locker room, champagne bottles were all over the place and everyone had left for the postgame party.

The craziest thing was how Darcie and the Tampa mayor, Jane Castor, put that boat parade together so fast. They were thinking about Friday or Saturday, but I told them the team's not gonna be here. When the league said we couldn't have a regular parade, they had the answer. They said we're going out Wednesday on the Hillsborough River. That's when Tom told us he'd stick around for the celebration.

Tom was feeling no pain that day. He had a hell of a time and the fans loved him for it. Brady became one of us that day. He got to be himself.

Now, we face the challenge of trying to repeat. It hasn't happened in more than fifteen years for good reason. When you win it all, you play so long and the other teams are all working to beat you. You've got to find ways to get to August fresh, both mentally and physically. That championship was won by last year's team. I'm gonna beat them in the face with that. All these banners going up around town? You guys didn't do it, they did it. You normally lose four or five good players in the offseason, but we didn't lose any starters, so that's a plus.

What helps is our locker room knows Brady has already done it. He knows what it takes to repeat, and he's not going to accept any crap on the practice field. That garbage stops right then and there.

Our first goal this season is to win our division because home games will be more important. We have to get that home-field advantage so our fans can go crazy. Everyone in the building knows what we're up against. We'll be on prime time five times this season, but hey, that comes with the territory. We've earned all the attention.

We're the defending Super Bowl champs, and we're proud of it. I can't wait to get started all over again. I know you'll love this book by Lars Anderson, reliving a season none of us will ever forget.

And always remember, both in football and in life: enjoy the journey.

I sure as hell have.

||||||||||||||||||||||||||||||||||||||||||||||||||||||||||||||||||||||||||||||||||||||||||||||||||||||

Tampa Bay Scheming

They dubbed it Operation Shoeless Joe Jackson. That was the code name Tampa Bay Buccaneer personnel tagged the team's pursuit of Tom Brady in the final days of winter 2020. It was a fitting moniker, because the odds of luring Brady to Tampa Bay—a team that hadn't been to the playoffs in thirteen years, a team with a .387 all-time winning percentage—seemed as likely to occur as a White Sox baseball player from 1920 emerging out of an Iowa cornfield in the movie *Field of Dreams*.

John Spytek, the Bucs' director of player personnel, had coined the classified name. He didn't want anyone to know that Tampa Bay was targeting Brady, who Spytek had played with for one season at Michigan, and so whenever the front-office heavies were talking about their aim to land the free agent quarterback with six Super Bowl rings, the code words were used. In the NFL, in matters large and small, secrecy is sacrosanct.

Spytek told other members of the Bucs' front-office staff that the one thing that really stood out when he played with Brady at Michigan was that the quarterback had shown frustration and disappointment only once during his senior season. It came after he threw an interception against archrival Ohio State. Brady slammed both hands onto his helmet and looked upward into the cold Midwestern sky for several moments, angry with himself for the blunder he had just committed. Brady was saved because the interception was wiped out by a Buckeye penalty, but Spytek's point was instructive: Brady was as consistent in his demeanor and approach to the game as any player he had ever been around in college—the kind of consistency at the quarterback position that the Bucs had lacked in 2019 with Jameis Winston. "And the Tom Brady we see now twenty years later is that same Tom Brady," Spytek told Buc staffers. "Nothing has changed about how he goes about his preparation and how he reacts to adversity."

Based on intuition alone, Spytek suspected that his former college teammate was looking for a new challenge. So he repeatedly told Bucs general manager Jason Licht, "If we build it, he will come," referencing the famous line from *Field of Dreams*. Spytek would then point to the depth chart posted on the wall in Licht's office at One Buc Place and mention that Tampa had a coach named Bruce Arians who had a long history of tutoring quarterbacks, and that Arians was as well-liked by players around the league as any head coach. It's no secret opposing players talk to each other in the NFL, and it wasn't uncommon to see players from other teams line up at the postgame tailgate party that Arians, a former bartender, would hold out of the back of his truck when he was the Arizona Cardinals head coach after games to give him a hug—even if they had never met him before. "To so many players in the league Bruce is like a cool uncle you want to have a beer with," said former Arizona quarterback Carson Palmer. "He's unlike any other head coach in the NFL in the way he relates to and treats players."

The underlying message Spytek was relaying to his general manager was clear. He believed Tampa had already "built" something that would appeal to Brady: a deep and dangerous group of wide receivers, a talented young defense, and a coaching staff that was as laid back as a Sunday-afternoon drive in the countryside—the opposite of the ultra-intense, my-way-or-the-highway Bill Belichick. Spytek also stressed that he genuinely thought Brady was ready for a fresh start, that he was going to finally break away from New England and emerge from under Belichick's shadow.

Still, Licht remained skeptical that Brady would actually leave New England. Licht had spent six seasons with the Patriots as a scout and front-office executive. He was sitting a few feet away from Belichick when the Patriots selected Brady in the sixth round of the 2000 draft. Licht believed—as did the majority of top football personnel around the league—that Brady would ultimately re-sign with New England and finish his career wearing only one team's colors. Wouldn't he?

But Arians made it clear: the Bucs needed to press forward with their clandestine operation.

|||||||||||||

Several weeks before the opening of the free agency period in March 2020, every team in the NFL was given a list of the free agent players who would be available at each position. The players were listed in alphabetical order. When Tampa quarterback coach Clyde Christensen examined the roster of free agent quarterbacks, he sat down at his desk to scroll through the list. Three last names beginning with the letter "B" popped out at him: Brady, Brees, and Bridgewater.

"In all my years of coaching I'd never seen Brady on the list,"

Christensen said. "Like basically everyone else in the league, I thought there was no way he would leave New England after twenty years. His life was there. He had roots planted there. His legacy was there. I thought that maybe he'd look around a little, but it would be nothing serious and he'd end up re-signing with the Patriots. The same type of situation existed with Drew Brees. I never thought he'd leave New Orleans. So initially we really zeroed in on Teddy Bridgewater. He was going to be our primary target."

Bridgewater had started five games for the Saints in 2019, completing 67.9 percent of his passes, throwing nine touchdowns and only two interceptions, and compiling a career-high 99.1 passer rating. Arians loved the fact that Bridgewater rarely threw interceptions: the anti-Winston. Bridgewater was from Miami, Florida, which meant he would be used to playing in hot and humid conditions. And he was an accurate passer, a skill that Arians has long believed can't be taught once a quarterback has reached his sophomore year or so of college. "You either have accuracy as a quarterback by the time you're about twenty years old or you don't," Arians said. "It's that simple. Sure, you can improve a little on it, but not much. Accuracy is the number one most important physical skill a quarterback must have to me. It's much more important than arm strength."

In February 2020, as more and more media reports out of Boston indicated that Brady may have thrown his last pass as a Patriot, Arians instructed Christensen to start studying Brady. "Tell me if he's still got it," Arians said.

Christensen had watched Brady for years. From 2002 to 2015, Christensen had served in various positions on the staff of the Colts. During that time, Indianapolis faced the Patriots fifteen times. Most of the games were close—nine were decided by seven points or less—but Brady had a knack for making one more play than did the Colts quarterbacks Peyton Manning and Andrew Luck: New England won ten of those fifteen games, including three of four in the playoffs. Whether from the sideline or the coaches box, Christensen—who mentored both Manning and Luck—marveled at Brady's ability to calmly deliver a game-changing pass when the stakes were highest. So many times, Christensen thought the Colts had delivered a knockout blow to the Patriots, only to have Brady methodically pick apart the Indianapolis defense and marshal a late-game, soul-crushing drive.

Before every kickoff against the Patriots, Christensen would be on the field and he'd see Brady go through his pregame routine of running up and down the sideline from goal line to goal line to warm up. Often Brady would pass within two feet of Christensen. It's commonplace in the NFL for quarterback coaches and coordinators to speak with the opposing team's starting quarterback before a game, but not once did Christensen approach Brady. He was always locked into his pregame routine, seemingly oblivious to his surroundings, his face frozen in a scowl.

"New England was never known for being a sociable organization, from the coaches down to the players, and I didn't want to bother Tom," Christensen said. "There were times when I'd tell myself, 'Okay, this is going to be the year that I'm going to do it, I'm going to say hi to Tom.' But then I'd get down on the field and he'd have that game face on. Man, he was so serious it was almost scary. So I'd always just let him do his thing and not bother him. But it just reinforced to me that New England was this dark, dark, dark place. They were almost nonhuman, like they were programmed to do one thing: kick your butt in football."

After Arians gave Christensen his assignment, Christensen spent most of his waking hours the next few days in his second-floor office at One Buc Place watching game tape of Brady. Deep into several nights, his head buried under the amber glow of lamplight, his face pressed close to his computer screen, Christensen analyzed every throw that Brady had made over the previous four years. He examined his arm strength, his accuracy, his ball placement, his drops, his footwork, his decision-making, his read progressions, his ability to keep his eyes trained down the field when oncoming rushers were closing in, his willingness to take a hit, his durability, and his toughness. Christensen would constantly pause plays to determine what Brady was seeing, then hit play and make a note of whether or not Brady made the right decision based on the defensive coverage he was facing and the offensive protection the Patriots had in place. He'd then rewind and watch the play again and again to ensure he wasn't missing anything. It was a painstaking, mentally exhausting exercise, but once it was over Christensen was certain in his assessment.

He walked into Arians's office to share his findings. "Tom has still got it," Christensen said. "He hasn't lost anything. The talent around him just wasn't as good last season. But in terms of his arm skill, there has been no drop-off. None. The guy can still do everything. He's still at the top of the class. I know it. You can watch the cutups of him for yourself and you'll see it, too. He hasn't lost a thing."

Arians can often be a man of few words, but now he was a man of no words. He simply nodded his head. Christensen returned to his office, unsure of what to make of Arians's silence. Arians then visited with Licht, who had conducted his own study on Brady. He acknowledged that the forty-two-year-old Brady wasn't in his prime, but Licht believed he still could make throws outside the numbers, and he had plenty of zip on his fastball. He added that his arm was far superior to that of Peyton Manning's when

Manning quarterbacked the Denver Broncos to a Super Bowl victory in 2016, even though Manning's arm was so shot he had trouble throwing the ball with force more than thirty yards down the field.

A few hours later, Arians stood at Christensen's office doorway. "I think we got a shot at getting Brady," Arians told his quarterbacks coach without a hint of emotion in his voice. Not uttering another word, Arians walked away. He had work to do.

|||||||||||||||||||||||||||||||||||||||||||||||||||||||||||||||||||||||||||||||||||||||||||||||||||||||||||||||||||

CHAPTER 2

Endings and Beginnings

They trudged off the field at Raymond James Stadium about twenty feet away from each other, the quarterback and the head coach, both in disbelief. The final play of the Buccaneers' 2019 season was a classic coach's nightmare: quarterback Jameis Winston threw a pick-six against the Atlanta Falcons on the first play of overtime—his thirtieth interception of the season, the most by an NFL quarterback in the twenty-first century. Tampa lost 28–22 in the shortest overtime game in NFL history: seven seconds. Arians knew—as did everyone in the Tampa organization—that this moment on December 29, 2019, and this play in particular, would mark the end of something.

Entering the final two weeks of the 2019 season, Tampa Bay's record stood at 7–7. Arians believed he had as much talent on his roster as any team he'd coached, yet the entire franchise was being held back by the play of its quarterback. Arians and offensive coordinator Byron Leftwich had repeatedly emphasized to

Winston before the final two games of the 2019 season that he needed to show the entire organization he was capable of being a Super Bowl–winning quarterback. "Jameis obviously was having a problem taking care of the ball," Arians said. "He wasn't processing information fast enough, wasn't seeing what he needed to be seeing on the field, and this slowed him down when getting the ball out of his hands. The throws were late and the defenders were getting their hands on the ball. Can't do that in this league and win a Super Bowl."

If the final two games were Winston's last chance to prove he could start for the Bucs in 2020, he failed miserably. At home against the Texans on December 21, he threw four interceptions, including a pick-six on the second play of the game, in Tampa's 30–23 loss. Then seven days later, he tossed two interceptions against the Falcons. The final one turned out to be the dagger to the heart of his career in Tampa, and it underscored his two fundamental deficiencies that were preventing him from becoming an upper-echelon NFL quarterback: his inability to read opposing defenses quickly and his what-the-heck-was-he-thinking decision making.

Here was the final play: With the ball on the Bucs twenty-five-yard line, facing first and ten, Winston dropped back to pass, looked to his right, zeroing in on tight end Cameron Brate. Facing no pressure, Winston held the ball one beat, two beats, then rifled a pass to Brate on a simple five-yard curl pattern. Falcons linebacker Deion Jones, reading Winston's eyes, jumped the route, picked off the pass, and ran untouched into the end zone for the game-winning, season-ending score. On the sideline, Arians threw his hands up in frustration, shaking his head, shouting expletives. Throughout the entire season Arians had consistently defended his quarterback. Arians had chalked Winston's various mistakes up to poor protection, or an incorrectly run route. No more. Speaking to the media fifteen minutes after that final play of the 2019 season, Arians called it like he saw it, saying Winston does "so much good and so much outright terrible."

Winston had hoped to be Tampa's version of Tom Brady after general manager Jason Licht had selected him with the number one overall pick in the 2015 draft. When Winston met Brady after the Bucs lost to New England, 19–15, in a Thursday-night game in 2015, Winston called it one of the thrills of his life. "I dream to be able to be the type of quarterback he is," Winston said. But he wasn't. In 2019 there was the good Winston who led the league with 5,109 passing yards. But there was also the terrible Winston who set an NFL record by throwing seven interceptions that were returned for touchdowns.

Arians didn't tell anyone outside the organization, but he wasn't sure he wanted to endure another season with his all-or-nothing, boom-or-bust quarterback. For his team to be transformed into a Super Bowl contender, he needed to at least try to find a new signal caller. "We need to look to see what is behind Door Number Two, or even if there is a Door Number Two," Arians told Licht. "We need to check out who is going to be available on the free agent market. I'm too damn old to groom a rookie. We're so close to being a great team. We just need one more guy."

That conclusion wasn't an easy one to draw for Arians. His loyalty to his players is legion, and he has a genuine affection for Winston. When Arians took the Tampa job before the 2019 season, he thought he could turn Winston into a Pro Bowl quarterback. In his exit meetings at the end of the 2019 season, Arians asked each of his top veteran players the same question: *Can we win with Jameis?* Every player responded with the same answer: *Yes.* But Arians was adamant that the Bucs needed to at least investigate the possibility of upgrading the position. The only way Arians would have Winston back in 2020 was if he couldn't find a seasoned free agent quarterback to replace him.

"B.A. felt we needed a change in culture, and that started with a change at the quarterback position," said Christensen, Tampa's quarterback coach. "As a person, we all loved Jameis. He worked his tail off for us. He was a good leader, a guy who was first in the building in the morning and the last to leave at night. He cared about football and he cared about Bruce. He didn't want to disappoint Bruce. But at some point you just have to realize when something isn't working."

Still, after the season, Christensen repeatedly asked Arians, "Are you sure you want to do this? Are you really sure we want to move on from Jameis? We're not even sure who we can get to replace him." Arians never directly replied to the questions, but Christensen had known Arians for four decades and he could read the body language of his head coach: *Yes, he was sure.*

|||||||||||||

At the NFL Combine in Indianapolis in late February 2020, Arians sent a not-so-subtle message to Brady and his team of advisors. A reporter asked Arians, standing at a podium inside Lucas Oil Field Stadium, during a press conference: *Of all the free agent quarterbacks, which one would you pick up the phone to recruit?* Normally a head coach would duck this question, telling the room that no decisions by the organization have been made and that all options are still under discussion. The last thing most coaches want to do is tip their hand about what players they are targeting, especially at the quarterback position. Plus, a coach never wants to be accused of tampering by publicly talking about a player who isn't on his roster. But the truth-telling Arians now used the microphone in front of him to make sure his voice carried nearly a thousand miles eastward to Boston and penetrated a gated mansion in the tony suburb of Chestnut Hill. Without missing a beat, Arians half-jokingly replied, "Tom Brady."

"You can't hit a home run unless you're going to swing for one," Arians later said. "You can't do anything special sitting on a fence."

At his home outside of Boston, as the start of the free agency period approached in early March, Brady studied the rosters of the four teams that interested him the most: the Chargers, the 49ers, the Bears, and the Bucs. He wrote down a list of twenty items that were important to him. He wanted to be close to his oldest son, Jack, thirteen at the time, who lived in New York. He wanted good receivers, much better than those he had had in New England. He wanted a strong coaching staff. And he wanted somewhere warm.

On the first day of the free agent negotiating period—it started at noon on Monday, March 16—Licht sat in his office at One Buc Place and phoned Brady's longtime agent Don Yee at precisely fifteen seconds after 12:00. "I'm calling about Tom," Licht said.

"You made the right call," Yee told Licht. "You really made a good decision to call me." Yee went on to explain that Brady had been paying close attention to the Bucs and Arians. He emphasized how much Brady respected Arians—who had written a book in 2017 called *The Quarterback Whisperer*—and Yee noted that Brady had been impressed with the work Arians had done with quarterbacks through the years. Brady had researched Arians, watching a documentary on him, and he admired how close Arians had been with all of his past quarterbacks. "You've got a good nucleus of talent at Tampa," Yee told Licht, "and it's important that the head coach, general manager, ownership, and the quarterback have the same commitment to winning."

Licht drove to Arians's house for a call with Brady on March 18, the first day the new league year opened at 4:00 p.m. EST. Licht arrived around 3:00 p.m. The two former bartenders "got some personality inside of them," according to Licht, before the call. Then, at exactly ten seconds after 4:00, Licht phoned Brady just as free agency officially began.

"What's up, babe?" Brady said as soon as he answered his cell. "Jason, this is going to be a hell of a lot of fun." Brady then gushed about the Bucs' talented wide receivers, the way the Tampa Bay defense had played during the second half of the 2019 season, and the way Arians had coaxed great seasons out of so many quarterbacks during his career. At one point Brady said, "I think we've got something. We've got a chance to be very special." Brady then told Licht why it would make sense for the Bucs to sign him.

After Brady uttered those words, Licht locked eyes with Arians, who was sitting two feet away in his kitchen, and flashed the closed fist thumps-up sign. Licht silently mouthed: *I think this is happening.* Like Captain Ahab, they figured they had their great white whale on the hook.

After thirty minutes of talking to Brady, Licht handed the phone to Arians. "If you come here, we will win the Super Bowl," Arians said. "You're the missing ingredient. We're a very talented team, but they just don't know it."

Arians kept talking, trying to pitch Brady on the virtues of playing in Florida for the Bucs. "Hey, Florida's great, it's warm, no state taxes. And we do have receivers Mike Evans, Chris Godwin, all these guys."

Brady interrupted, saying, "Hey, I think Gronk might want to come out of retirement."

"Let's get you signed up first and then we'll work on that one," Arians replied.

Brady then took command of the conversation, repeating what he liked about Tampa. While he and Arians spoke, Licht paced back and forth in the kitchen like he was on some kind of guard duty. As the minutes passed and the smile on Arians's face grew, Licht understood what it meant: they had their man.

Arians eventually told Brady he'd talk to him soon and then handed the phone back to Licht. "Hey, there's one other thing here," Licht said. "It's a small thing, but maybe a big thing. We have a number 12 on our team and he's pretty good—Chris Godwin. What are you thinking about that?"

"Oh, he's a great player," Brady replied. "I'm not going to take his number. I don't care about that. You know what number I'm thinking of? I'm thinking of taking maybe number 7. Is that available?"

"It is," Licht said. "Why do you want that number?"

"I want that seventh ring," Brady said.

NFL players have taken extreme measures in the past to secure their favorite jersey number. Back in 1995 Deion Sanders of the Dallas Cowboys bought a BMW for Alundis Brice in exchange for number 21, which had been Brice's number. The Washington Redskins' Clinton Portis paid $38,000 to Ifeanyi Ohalete for the privilege to don the number 26 jersey. And in 2014 Tennessee Titans backup quarterback Charlie Whitehurst, who had worn number 6 since his rookie season of 2006 with the Chargers, arm-wrestled punter Brett Kern for the right to wear number 6— and lost.

But Godwin had other ideas. A few days before Brady had talked to Licht about the number 12, as rumors swirled in the Tampa facility that the team may sign Brady, Godwin was lifting weights when he told teammates, "If we get Tom Brady, I'll gladly give up number 12." When the two finally spoke, Brady told the receiver that he had no problem wearing number 7. But Godwin, wanting to start his relationship with Brady on a positive note, eventually told Brady he'd change his number to 14 and let him wear 12 because of the "respect" he had for his new quarterback. Like any wise wide receiver—especially one entering the final year of his contract—Godwin wanted to win over the man who would be in charge of how many throws came in his direction.

|||||||||||||||

After their call with Brady ended, Licht, Arians, and Arians's wife, Chris, enjoyed a celebratory dinner at Ava, an Italian restaurant on Howard Avenue. Licht shared another nugget of information from his talk with Brady: The quarterback asked for the cell phone numbers of all the Tampa players. "Damn, that shows his commitment right there," Arians said. "He hasn't even signed his contract yet, and he's already thinking about how he's going to connect with his teammates. This is what leadership looks like. He's going to make us better the second he walks into the building, even before he makes his first throw."

During dinner at Ava, Arians spotted Raheem Morris sitting at another table. Arians waved Morris—the former head coach of the Buccaneers from 2009 to 2011—over. "What are you guys celebrating?" Morris asked.

"Oh nothing," Arians said, concealing his joy. "Just out enjoying a good meal."

After more small talk, Morris walked away. Only three people in the state of Florida knew the biggest secret in sports.

||||||||||||||

In retrospect, the fracturing of Brady's relationship with Belichick and the Patriots could be seen for years. Each time a high-profile player was cut from the New England organization—the list includes linebacker Willie McGinest (who was selected to play in two Pro Bowls), running back Corey Dillon (four Pro Bowls), safety Lawyer Milloy (four Pro Bowls), and cornerback Ty Law (five Pro Bowls), to name a few—Brady would stand at a podium at One Patriot Way, summon a half-grin, and then tell the assembled media in a monotone voice that he had nothing of substance to say about the subject. But his dejected body language and manufactured smile always suggested displeasure.

As successful as it was, the Patriot Way—the guiding principle of the New England organization—could be bloodless and soul-crushing. What is the Patriot Way? Bill Belichick has claimed that he has never uttered that term (he once excoriated a reporter for implying he did), but it essentially means that the New England franchise is all business, all the time. Winning is the only thing that matters in the Patriot Way, which explains why the organization isn't afraid to take chances on high-risk, high-reward players other teams refuse to pursue (see Randy Moss in 2007, LeGarrette Blount in 2014, and Antonio Brown in 2019).

At the same time, Belichick and Company won't hesitate to trade a Pro Bowl talent who isn't viewed as a team player (defensive tackle Richard Seymour in 2009 and linebacker Jamie Collins in 2016). They also won't shy away from shedding beloved former stars on the back end of their careers (linebacker Mike Vrabel in 2009) or those seeking a contract above their level of play (wide receiver Deion Branch in 2006). In the Patriot Way, the team always trumps the individual, and decisions to protect locker room chemistry and the organization's future are made swiftly and without emotion.

"I swear it seems like Bill has listening devices in their locker room," said a longtime NFL coach in 2017. "If there's a player with a bad attitude, he'll know about it and then that player will be gone. There is an element of fear on that team—the fear of getting shipped away to a place like Cleveland—but that keeps everyone in line. It keeps everyone going the Patriot Way. Everyone is expendable. Unless you're Bill Belichick or Tom Brady, you could be gone at any minute."

Yet for years, people around the league wondered whether the Patriot Way would ever make Brady, too, expendable. Brady's father once said that as soon as Belichick had a quarterback who was as talented as his son and was a buck cheaper, Tom would be shown the door. In a 2018 interview with Jim Gray, Brady was asked whether he felt appreciated by his bosses for all that he accomplished. "I plead the fifth," Brady said. A few close to Brady say he never felt fully respected by Belichick. Everyone wants to earn the admiration of their boss and—whether it was due to a communication breakdown or not—Brady simply didn't receive the positive reinforcement from his head coach that he was seeking, that he needed. During the 2019 season, after most Patriot victories, the ultra-fast embrace he typically exchanged with Belichick was the definition of awkward, like a couple hugging moments before signing divorce papers.

Brady's last pass as a Patriot was on January 2, 2020, in a 20–13 wildcard game loss to the Tennessee Titans. Facing a first and ten on his own one-yard line in the fourth quarter, Brady unleashed a ball that landed in the arms of Titans cornerback Logan Ryan, who returned the interception nine yards for a touchdown.

After the game it appeared Brady didn't even shower—a rarity for him—before he held a brief postgame press conference and rushed through the concrete bowels of Gillette Stadium with his daughter in his arms, clearly not wanting to be bothered. With his wife next to him, the Brady family strode into the rainy New England night, disappearing into the darkness. He would never be seen again at One Patriot Place.

About three months later, on the evening of March 16—two days before his call with Arians and Licht—Brady drove to Patriots owner Robert Kraft's mansion. The two had a friendly, extended discussion. Kraft believed that Brady had come over to work out a new contract, but no: it wound up being their final goodbye.

"Look, I just want to say I love you and I appreciate what we've done together," Brady told Kraft. "I know that we're not going to continue together, but thank you. Thank you for providing what you have for my family and my career."

Tears ran down Brady's face as he spoke. Sitting close to Kraft, he then pulled out his cell phone and called Belichick to tell him he was leaving the Patriots. It was important for Brady that his coach hear the news of his decision from him, not from a reporter or another secondary source. Brady wished his coach "the best" and, his eyes still wet, he thanked Belichick for all he had done for him over the previous two decades. Kraft later didn't directly say the reason Brady was leaving was Belichick, but he implied that it was. "Think about loving your wife and for whatever reason, there's something—her father or mother that makes life impossible for you and you have to move on," the owner told the NFL Network.

The next day, March 17, in a series of Instagram posts, Brady said he was moving out of New England. He thanked everyone from Kraft to Belichick to his teammates to the low-level staffers.

But he made it clear that—after three MVP awards, forty-one playoff starts, thirteen conference championship appearances, four Super Bowl MVP trophies, six Super Bowl rings—that he was ready for a change.

So why did Brady ultimately leave? Even he has trouble articulating his complicated, nuanced, often challenging relationship with Belichick—the two would never be mistaken for regular dinner-party companions. But one fact is clear: Brady wanted to play until he was at least forty-five years old, and dating back to 2017, Belichick would not give him a contract extension. As a result, Brady didn't feel that Belichick truly valued him, which in turn made football a job for Brady, not a joy. He was searching for fun again. He was searching, in the end, for a season in the sun.

||||||||||||||

On April 16, 2000, Jason Licht was in New England's Draft War Room with Belichick and other front-office personnel, trying to get the most out of the Patriots' final picks. On the other side of the country, in the living room of his family house in San Mateo, California, a young Tom Brady sat with his family riveted to the television, watching the drama unfold from Madison Square Garden.

The Brady family had hoped that their Tommy would be selected by the San Francisco 49ers. The Bradys had been season ticket holders since before Tom could walk, and Tom Sr. had taken his only son to his first Niners game when he was four years old. On that day, January 10, 1982, San Francisco beat Dallas in the NFC Championship after Joe Montana connected with Dwight Clark late in the fourth quarter on "The Catch," one of the most famous plays in NFL history. Ever since, Tom Sr. dreamed that his boy would play for San Francisco.

The Niners had four draft picks in the first two rounds—and used them all on defensive players. In the third round San Francisco picked a quarterback: a player named Giovanni Carmazzi from Hofstra, prompting one of Brady's sisters to ask, "Who's he?" When the fourth round ended and Tommy still hadn't been picked, worry permeated the Brady household. Was it possible he wouldn't be drafted at all?

In the fifth round the Brady clan was certain Cleveland would take him. Dwight Clark—the same man who little Tommy had watched so many years ago at his first NFL game—was the Browns' director of football operations and Cleveland needed a quarterback. Clark was familiar with Brady. He knew Brady had repeatedly come off the bench at Michigan and rallied his team to victory. He had heard the Wolverine coaches describe Brady as "a hard worker" who was determined to succeed in the NFL—at any cost. But then the pick was announced from the podium in New York: *The Cleveland Browns select quarterback Spergon Wynn from Texas State University.*

Tom was devastated. "I don't understand this," Brady said. "I do *not* understand this."

"Dwight Clark—unbelievable," said Tom Brady Sr.

"I gotta get out of here," Tom said. He went upstairs to his bedroom, dug a baseball bat out of the closet, and went outside. After a few minutes, father and mother chased after him. They found their son crying. Putting their arms around him, they went for a walk. "Basically they're saying that I don't look like an NFL quarterback," Tom told his parents as they meandered through their neighborhood.

Back in Foxborough, the Patriots had no need for a quarterback. Drew Bledsoe was firmly entrenched as the franchise signal caller. With the 187th pick in the sixth round, New England picked Antwan Harris, a cornerback from Virginia. But Belichick kept eyeing the draft board, wondering why Brady hadn't been selected. He kept saying that Brady would be a value pick at this point, even if he never became a high-end NFL starter. Future All-Pro quarterbacks simply don't go undrafted until the sixth round, but still Belichick didn't understand what he was missing. He also didn't know why the coaches at Michigan—where Brady had split time with Drew Henson for much of his senior season—didn't seem to believe in him.

But Belichick went with his gut and his eyes. The Patriots had the 199th overall pick of the sixth round, and he told his personal assistant, Berj Najarian, to call Brady. When the phone rang in San Mateo, Tom was still outside, trying to process all that was happening—or, more accurately, *wasn't* happening. Tom Sr. answered the phone and then hustled to get Tom.

"Hello," Brady said.

"Tom, this is Berj Najarian with the New England Patriots. We're about to draft you. I'll pass you over to Bill."

"Good to have you," Belichick said before Brady could say a word. "Looking forward to getting you here. We're gonna work hard. See you soon."

And that was it. Tom Sr. uncorked a bottle of Champagne. The announcement was made from the podium at Madison Square Garden: "With the one hundred ninety-ninth pick in the draft, the New England Patriots select Tom Brady." From the back of the draft War Room in Foxborough, Licht watched it all.

A few days after the 2000 draft, twenty-two-year-old Brady flew from California to Ann Arbor, packed up his apartment, then drove, alone, for eleven hours to Boston. One question weighed on his mind more than any other: What would it be like to play for Bill Belichick?

|||||||||||||

After speaking with Arians and Licht on March 14 and then conferring with his agent Don Yee, the quarterback privately made up his mind: he was signing with Tampa Bay. He and his wife, Gisele Bündchen, had already decided they were going to relocate their primary residence to Florida, and even though Gisele had expressed concern about her husband continuing to play football into his forties, she gave her blessing for the move to Tampa.

Brady's marriage was in a good place. Two years earlier, he had missed the voluntary sessions of New England's offseason training program, which was interpreted by some in the Boston media as another example of the widening gulf between Brady and Belichick. But in truth, Brady wanted to work on his marriage and spend more time at home with Gisele, a supermodel from southern Brazil whose career earnings as of 2019 were $488 million, or more than twice what her husband's on-field earnings had been to that point ($235 million). During that time Gisele had written a letter to her husband that detailed her issues with their marriage and said that he needed to be more active in taking care of family duties. After reading the missive, Brady shined a spotlight on his soul, examining his priorities, looking at how much time he was spending with their two young children and his son, who lived with his former girlfriend in Manhattan.

The couple went to marriage counseling, working through their problems, which were both small (Brady didn't drive the kids to school enough) and big (Gisele felt like she didn't have the time to pursue her own professional goals). They talked, they listened, and they grew closer together, remembering why they fell in love in the first place, shortly after their first blind date at Turks & Frogs, a Turkish wine bar in New York City's West

Village. Brady eventually placed the letter Gisele had written to him in a drawer, but he'd pull it out on random nights and reread every word, wanting to be reminded that he always needed to tend to his marriage with the same devotion and tenacity that he applied to his day job.

Now they were ready to begin another chapter together, in a new place. After discussing it with his wife, Brady called his agent Don Yee and told him to broker the deal with the Bucs. Licht and Yee quickly came to terms on a two-year, $50 million contract. On the morning of the agreement, Brady was in his Manhattan penthouse apartment in the upscale Tribeca neighborhood: a five-bedroom, five-and-a-half-bath spread that stretched for 4,567 square feet and occupied the entire twelfth floor at 70 Vestry. It included a library, a billiards room, a steam and sauna room, hot and cold plunge pools, and a private courtyard. The couple would eventually sell the apartment in January 2021, for $36.8 million.

There was just one problem that day: Licht and the Tampa owners insisted that Brady take a physical before signing the contract. Licht had hoped to have Tampa's own medical staff examine Brady, but because of COVID-19, the team couldn't get Brady to Florida. Instead, the Bucs team doctor found a physician in New York for Brady to visit.

Brady was a little miffed at the directive. "The only major time I've missed in the NFL in twenty years was in 2008 when I hurt my knee," Brady told Licht. "Trust me, I'm good to go."

And he was: Brady passed the physical. As soon as Licht received the positive medical report, he was confronted with another pandemic problem: How would he get Brady to formally sign the contract? Under normal circumstances, Brady would have flown to Tampa for an individual meeting with the Bucs ownership during which he'd sign the contract and then he'd hold a press conference with Licht and Arians. But now? Licht had to get creative.

Mike Greenberg, who was Tampa Bay's director of personnel administration and in charge of managing the Bucs' salary cap, had a brother-in-law who lived in New York City. Greenberg scanned a copy of the contract and emailed it to his brother-in-law, who then printed it out. The instructions Greenberg gave his brother-in-law were simple: get in a cab, go to Brady's apartment, and have him sign it.

The brother-in-law did as he was told, riding in the back of a taxi through the streets of Manhattan, which were suddenly empty because of the outbreak of the coronavirus. In the car, the brother-in-law's cell rang; it was Licht. "Don't fuck this up," Licht said. "Wear a mask. Don't shake his hand."

"I got this," the brother-in-law replied.

The driver pulled up to Brady's building. Carrying the contract in a shoulder bag, wearing a mask, the brother-in-law stepped into the lobby, where a doorman stopped him. The brother-in-law said, "I'm here to see Tom . . ."

"I know," the doorman said. "Tom's waiting for you."

"Where is he?"

"Just press the twelve button on the elevator."

"What apartment number is it?"

"Just press twelve on the elevator. You'll see."

A mess of nerves, the brother-in-law rode the elevator to the twelfth floor. The doors opened, and spread out before the brother-in-law was the entire apartment. He was already inside. Not sure what to do, he said, "Hello? Hello? Tom?"

"I'm in here," a voice yelled from the distance. "I'm in the kitchen. Just follow my voice."

Stepping into the vast penthouse, the brother-in-law finally found Brady. He put the contract on the kitchen counter. As Brady signed it, the brother-in-law, holding up his cell phone, snapped a picture, which he then texted to an employee in the Bucs public relations department. And just like that, the team had bagged its quarry, its great white whale. Tampa's secret stalking of Tom Edward Patrick Brady Jr. had ended in success.

|||||||||||||

On March 20 at 9:31 a.m., the Tampa Bay social media department tweeted out the photo of Brady standing in his kitchen in his New York City apartment, pen in hand, signing his contract with the Bucs. He wore a black hoodie and an unmistakable look of excitement. He had his new challenge.

Moments after Brady had put pen to paper, he was on the phone with Licht. No detail is too small or inconsequential for Brady, who had already done the math on how much time he had to get ready for the season opener: He informed Licht they had roughly 4,270 hours for him to learn the Bucs offense and the quirks and likes of his new wide receivers and tight ends before the regular season kicked off. "I've got to get to work," Brady told his new general manager. "I've got to get started right now."

Brady then reminded Licht that he was now in a division with New Orleans, where quarterback Drew Brees and head coach Sean Payton had been together for thirteen seasons. Also in the NFC South were the Atlanta Falcons, whose quarterback (Matt Ryan) and coach (Dan Quinn) were about to start their seventh year together. Brady noted that either the Saints or Falcons had finished ahead of the Bucs in the final regular season divisional standings in fifteen of the last seventeen years, and if Tampa Bay was going to bust this trend, he had to catch up with Brees and Ryan in terms of his comfort-level with the offense. Brady reiterated to Licht, "Our season starts now. I've got a whole new language to learn in the playbook. These other guys in the division don't."

As Brady and Licht chatted, Arians was playing golf at Tampa's Old Memorial Golf Club. For days rumors had circulated among the regular patrons at Old Memorial that the great Tom Brady might be coming to Tampa. The last time the team had a shot at a franchise-changing quarterback was in 1984; some old-timers still remembered the day the Bucs picked Steve Young first overall in the NFL Supplemental Draft of USFL and CFL players. Alas, Young had little desire to play for the team and the future Hall of Fame quarterback was traded a few years later to San Francisco. But now on this spring morning, the scuttlebutt among the weekend duffers was picking up steam: Brady to Tampa *seemed* like a real possibility.

A few of the regulars at Old Memorial even mentioned this to Bucs quarterback coach Clyde Christensen before he teed off with Arians. The two were playing a friendly match and at the ninth hole—a 550-yard par five—Arians duck-hooked his T-shot into the rough. Then Christensen pulled out his driver and sent a 280-yard bomb straight down the fairway. The two were tied through eight holes, and now Christensen had the advantage in their nine-hole five-dollar bet.

From the tall grass, Arians blasted a three-wood, propelling his ball to 120 yards from the green. After driving his golf cart to his ball, he grabbed a pitching wedge. As he lined up his third shot, a golf cart charged at him from behind. "Bruce, Bruce!" yelled a voice. It was Licht. "I've got someone on the phone you need to talk to." Arians put down his club and Licht handed him his cell phone.

"I'm a Buccaneer," Brady said. "Let's get to fucking work. Let's fucking do this."

"Hell yeah, baby," Arians said. "Everyone is going to follow your lead." Coach and quarterback, now officially united, talked for about ten minutes. They were already mapping out their plan to win a Super Bowl. Brady would start calling his teammates as soon as he got off the phone. He wanted to forge a connection with every single player he didn't know, which was essentially the entire roster. Arians was again blown away by Brady's humility and commitment to winning just minutes after signing his contract. This was an example of what Arians calls "grit"—in Arians's definition of "grit," the word is synonymous with character. To Arians, this quality is the number one factor that distinguishes the great quarterbacks from the average ones.

He believes that 95 percent of the successful quarterbacks in the NFL are special *people*, not just robots with big arms. They inspire others. They get teammates to do things they never thought possible. Arians has coached a few quarterbacks in his career with off-the-charts grit—quarterbacks that have rare mental skills and a unique strength of character—and after just two conversations with Brady he was sure that he had those qualities as well. "Once you cross paths with someone like this—whether it's in football or academia or politics—that person will stick with you for the rest of your life," Arians said. "Right away with Tom, I knew he was as gritty as they come. Nothing stops guys like this. Nothing."

Arians told Brady to hang on a for second—he needed to wrap up his front nine. He then picked up his pitching wedge again and stroked the ball cleanly, sending it high into the soft blue Florida sky. The ball landed five feet from the hole and rolled to within two feet. Arians smiled wickedly at Licht. Already, the coach's fortunes were improving now that Brady was with him.

He drained the putt for a birdie; Christensen made a par. Arians had won the hole and the front-nine bet. As he walked off the green, Licht handed him the phone back.

"Fucking birdie, baby," Arians told Brady. "I just beat your quarterback coach's ass. We're off to a good start, baby. A fine start."

For Arians, the number one thing Brady would bring to Tampa Bay was an intangible: confidence. "I knew it would legitimize our locker room very much like Carson Palmer did when he came to Arizona," Arians said. "Confidence is the ultimate X-factor in the NFL. The difference between winning and losing is so tiny in this league. Games always come down to just a handful of plays, sometimes just two or three. And for years Tom had always delivered in the clutch. No moment has ever been too big for him. We knew we already had talent on our roster. We just needed belief. I knew Tom would give us the belief."

A few days after talking to his new coach on the golf course—and two decades after he drove halfway across the country to begin his career with the Patriots—Brady flew on a private jet to Tampa Bay. During all those hours behind the wheel in the spring of 2000 he wondered what his relationship with Bill Belichick would be like. Now a similar question dogged Brady as he looked out the window from 30,000 feet: What would it be like to play for Arians, who in so many ways was the anti-Belichick?

The private jet began its descent into the bright orangish Tampa sun, the blue waters of the bay glistening below. The journey had begun.

||||||||||||||||||||||||||||||||||||||||||||||||||||||||||||||||||||||||||||||||||||||||||||||||||||||

CHAPTER 3

The Arrival

He steered his black Tesla Model S through the streets of South Tampa Bay, a forty-two-year-old man on a mission. Outside the eco-friendly car's tinted windows, he could see the lush greenery of this coastal city and the gently swaying of cyprus and magnolia and crape myrtles in the warm Gulf Coast breeze. With every mile he drove, his anticipation grew. The date was April 7, 2020, and Tom Brady's new life in Florida was just beginning. He was yet another longtime northerner who had fled the gray chill of New England, and now he was on his way to pick up vital business information. As the black Tesla continued to roll through the springtime afternoon and approach his destination near MacDill Air Force Base, in a quiet neighborhood with manicured lawns, he listened to the directions voiced by his GPS. Like anyone in a new city, he didn't want to become lost.

Brady finally spotted a gray house with white trim. The small driveway led to a large, ground-to-roof window. The house next door had a nearly identical look, layout, and outward appearance of the one that Brady was now zeroing in on. He parked his Tesla, stepped out of the car, and walked toward the front door. Rather than meeting face-to-face at a different location, Brady and his new offensive coordinator Byron Leftwich decided to get together at Leftwich's house. The critical business materials Brady needed to retrieve included a three-ring binder that contained a hard copy of Tampa's playbook and an iPad that also had the playbook on it. The two agreed to wear masks and practice social distancing during their brief interaction. (The NFL had issued a work-from-home order on March 25, when Commissioner Roger Goodell told teams to shut down their football facilities to everyone except players receiving medical treatment and a handful of employees who were deemed essential workers.)

Brady approached the residence, his six-foot-four, 225-pound frame casting a long shadow in front of him. But it wasn't Leftwich's house. Inside, a man named David Kramer sat at his kitchen island, working on a laptop. Kramer had put his house on the market and was renovating certain rooms, so he left the door unlocked for construction workers, who were constantly coming and going. But now Kramer wasn't expecting any workers or guests, and he looked up from his computer and spotted the shadow of a very tall man on his porch. The mystery visitor wore a baseball hat and carried two duffel bags. A few heartbeats later, with Kramer's confusion growing, he saw the doorknob on the front door turn. Watching it move, Kramer was overwhelmed with one thought: *What the hell is going on here?*

Brady opened the door and stepped forward. Dumbfounded, Kramer sat in stunned silence as Brady entered his home. The two made eye contact. In a causal, ho-hum voice, Brady said, "How's it going man?" He put down his two duffel bags. Then he strode toward Kramer, who was somewhat put at ease by Brady's calm demeanor, the confidence of his gait, and the fact that he was a clean-cut, good-looking man—hardly the type of person who would appear to harbor any bad intentions. Brady acted as though he belonged there.

Finally gathering himself, Kramer replied to Brady's question. "I don't know," Kramer said in a sarcastic tone. "You tell me, dude."

Then Brady looked closer at Kramer, puzzlement spreading across his face. Kramer kept his eyes on Brady, and then it hit him as quickly as a star shoots across the midnight sky: standing before him was the NFL's most high-profile, most recognizable player.

"Am I in the wrong house?" Brady asked.

"I think so," Kramer said. "But who are you looking for? Where are you supposed to be?"

"Is this Byron's house?"

Brady froze. Kramer froze. Time ticked—one second, two, three—the pair staring at each other. "I am so sorry. I am so sorry," Brady said.

Holy cow, thought Kramer. *Tom Brady is in my house.*

Brady apologized again and again, quickly grabbed his bags, then bolted out the front door. Kramer, who had been working from home by himself for months and was often consumed with boredom, wanted to yell after him: *Tom! Tom! Tom!* But he was so astounded that the words wouldn't come out. Brady beelined it to Leftwich's house next door, carrying his two duffel bags that were filled with water bottles. Kramer didn't even have the wherewithal to snap a photo.

For Brady, it was his first Florida-man moment. It wouldn't be his last.

||||||||||||

Brady picked up both the three-ring binder and the iPad with the team playbook. "Tom is old-school when it comes to his playbook," Arians said. "He likes to see it on paper." This was during the NFL's "dead period," before the virtual offseason began. During this stretch in the spring players weren't allowed to enter the team facility or have any organized classroom instruction or conditioning. Once rival NFL teams found out that Brady had visited his offensive coordinator, a few complaints were voiced to the NFL commissioner's office in New York that the Brady-Leftwich interaction constituted in-person instruction, but league officials determined no rules violations had occurred.

At the time Florida was under a stay-at-home order, which advised Floridians to only travel for essential purposes to avoid contracting the coronavirus. Was the visit to Leftwich a violation of that order? This much was clear: Brady's first quarterback sneak as a Buc, though not perfect, was successful. And the episode had a happy ending for David Kramer as well. Kramer shared the anecdote with his realtor, who then began telling clients at showings that this was the house that the great Tom Brady once accidentally walked into. The home sold faster than Kramer ever dreamed.

⸻

Shortly after Brady signed with Tampa, Leftwich received a text message from Steelers quarterback Ben Roethlisberger, with whom Leftwich had played in Pittsburgh for five seasons. "Hey, don't screw it up," Roethlisberger texted. Leftwich replied: "All I got to do is get out of the way." Leftwich's role had suddenly changed. When he was coaching Jameis Winston in 2019, he repeatedly discussed with Winston what his quarterback was seeing on the field and where he should go with the ball based on the scheme the defense was playing. Brady would need no such guidance, and Leftwich planned to give him much more freedom and license with the offense.

The day Brady became a Buc, he and Leftwich spent more than an hour on the phone. "Just coach me," Brady said to his new offensive coordinator, a former first-round pick in 2003 by the Jacksonville Jaguars who also played for the Falcons, Steelers, and Buccaneers before retiring after the 2012 season. "I want to see how you see the game," Leftwich told Brady. "We're going to pick each other's brains and we're going to get you up to speed in this offense. It could take longer than we think."

Brady and Leftwich went way back. Leftwich first played against Brady in 2003, when Leftwich was a rookie for the Jaguars. From the sideline on a snowy December afternoon in Foxborough, Leftwich marveled at Brady's calmness on the field, noting that he was so unfazed by on-rushing defensive linemen that he often stood flat-footed in the pocket. "No one does that," Leftwich said. "That just showed how in control he was. He knew where everybody was on both teams all the time. He could just stand in the pocket and slide a little to avoid chaos. He displayed so much patience and everything just looked so instinctive and easy for him. Of course, it's not easy, but like all the great ones, he made it seem so simple." Brady threw for 228 yards and two touchdowns that day as New England won the game, 27–13.

They would face one another again in the 2005 playoffs. In the only playoff start of his NFL career, Leftwich completed 18 for 31 passes for 179 yards and one interception. Brady was a machine of efficiency in the Patriots' 28–3 win: he completed 15 of 27 for 201 yards and three touchdowns. After the game, the quarterbacks embraced near midfield, with Brady telling Leftwich that he was sure the two would meet again in the playoffs one day. That never happened, but now they were on the same team.

Brady established their relationship hierarchy during their first talk: he was the pupil; Leftwich the teacher. It didn't matter that Brady had a small mountain of Super Bowl rings. It didn't matter that Brady was two years older than Leftwich. He wanted to be coached and, more than anything, wanted to get better every single day.

Arians especially believed in the potential fruitfulness of the Brady-Leftwich relationship. Because they were so close in age, they knew many of the same people in the NFL and could even understand cultural references from the 1990s when they were in college—they could talk, for instance, about *The Fresh Prince of Bel-Air*—that many young NFL quarterbacks don't get. Also, Leftwich had played quarterback in the Arians offense at Pittsburgh, so he knew the progressions the quarterback needed to go through as defenders were grabbing at the quarterback's arms, legs, and jersey. Arians even believed that Leftwich would do a better job of teaching Brady than he could because Leftwich had been in the pocket in this system when the action was live; Arians had only coached it from the sideline.

"Tom and I had a high level of trust from the very start because we knew each other as players and he knew that I had run this offense as a player," Leftwich said. "We came from the same era of football, and we were drafted just a few years apart. We could relate to each other, which I think made our communication even better right out of the gate with me being his coach and him being our quarterback. And even from our first conversation, he made it very clear that he wanted to be coached and that he truly wanted me to coach him so he could get better, even in his forties."

During that first conversation, Leftwich was struck by Brady's humility and drive. And now, a little more than two weeks after signing his contract, Brady had his hands on the Bucs playbook—the start of the long, laborious, tedious process of learning the high-flying, risk-taking Bruce Arians offense, which was nothing like Brady orchestrated in New England the previous twenty years.

|||||||||||||

It was a call that every general manager and coach dreads: informing a player—especially a player who was once considered the face and the future of the franchise—that his time with the team was over. After Brady agreed to his deal, this wouldn't come as a shock to Jameis Winston. But it didn't make the task of cutting Winston any easier for Jason Licht. It had been Licht's decision to select Winston with the number one overall pick in the 2015 draft, and now it was Licht's responsibility to formally sever ties with him.

There was an emotional connection between Licht and Winston as well. Licht's three children—Charlie, Zoe, and Theo—were rabid Winston fans. "They have his jersey in their rooms," Licht said. "It made making that call all the more difficult. But the NFL is a tough business."

Licht phoned Winston. "Hey man, as you probably know, we're going in a different direction."

"You've signed the GOAT [greatest of all time]!" Winston said, almost sounding happy for his friend Licht.

"We did," Licht said. "But you're going to go on and do great things. You've given us everything you've got. You've worked your ass off. I'm going to tell everybody who asks me what a terrific person you are and what a great leader you are. Your future is bright."

"I appreciate you so much, Jason," Jameis replied. "You are the G.M. who picked me number one. I'll never forget that and I'm so grateful. Make sure you say hello to Blair [Licht's wife], Charlie, Zoe, and Theo for me."

"Remember," Licht said, "you still have a chance to be a hell of player in the league."

"I plan on it," Winston replied. "I plan on it."

Back in the early 1980s, Arians was on Alabama coach Bear Bryant's final staff. Bryant often counseled a young Arians, the Tide's running back coach, telling him how important it was to care for players off the field. Arians had the Bear in mind when he called Winston and thanked him for his hard work and emphasized how much he appreciated him as a person. It was a short conversation, but Arians conveyed a powerful message: in the future, if Winston ever needed anything—whether in football or in life—Arians was only a phone call away. Winston had attended the Arians Family Football Camp in Birmingham, Alabama, when he was in tenth grade, the beginning of Arians's mentorship of Winston. To this day, neither coach nor quarterback has ever uttered a negative word about the other.

||||||||||||

Brady and his family moved into an empty, Shangri-La–like estate on Tampa's Davis Island. They were just renting. Their landlord? That would be Derek Jeter, the former New York Yankees shortstop who was now a co-owner of the Miami Marlins. Jeter relocated to Miami in 2017 to be closer to the Marlins, and he was more than happy to turn the keys of Tampa's most palatial spread—it was the largest house in Hillsborough County—over to Brady, one of Jeter's good friends. But still, Jeter charged his tenant rent to live in "St. Jetersberg," as some locals called the property.

Jeter spent four years building the mansion. (He would put it on the market in July 2021 with a listing price of $29 million, and it sold for $22.5 million to an unidentified buyer.) The nearly 22,000-square-foot waterfront estate on Davis Island—which is only a short drive away from the Bucs facility—sat on a 1.25-acre lot. It has an eight-foot privacy fence that neighbors call "The Great Wall of Jeter." Outside, there's an 80-foot saltwater lap pool, a 90,000-square-foot waterfront patio and outdoor kitchen, a deep-water dock and two boat lifts—the ultimate party palace. Inside, the two-story main house features seven bedrooms, eight full baths, eight half-baths, and a gourmet kitchen with two marble islands, two Sub-Zero refrigerators, and four dishwashers. More indoor amenities include a gym, a movie theater, a wine cellar, and a wood-paneled office. There is an open foyer with 24-foot-high ceilings and floor-to-ceiling windows that provide spectacular views of Hillsborough Bay.

Almost as soon as Brady, Gisele, and their two children moved in, the mansion became a tourist attraction, a must-see stop for curious onlookers. At the family's old fortress in Chestnut Hill, outside Boston, the public couldn't get close enough to spot the famous quarterback and his supermodel wife. Here, before the boxes were even unpacked, boaters were circling outside the Bradys' dock, hoping to catch a glimpse of Tampa's newest residents and most famous couple.

There was another nuisance, too, in those first days in Tampa. One afternoon, Gisele was paddle boarding in the Bay when, suddenly, she spotted what she thought might be an alligator in the brackish water. And the creature was close. She made it back to the dock safely, but the scary episode—along with gawking fans—reaffirmed that she and her family certainly weren't in Massachusetts anymore.

||||||||||||

Ryan Jensen, the Bucs' starting center, was in bed in his offseason home in Evergreen, Colorado, when his cell phone rang at 6:45 a.m. on March 20. The caller ID read: Jason Licht. Jensen quickly popped up, cleared his throat, and answered. Was something wrong? Why was Licht calling him just as the sun was rising over the Rocky Mountains?

"We just signed Brady," Licht said, "and I wanted to give you a heads-up. He's going to call you in fifteen minutes. He's really excited to talk to you."

On schedule, Brady popped up on FaceTime. Not expecting a video interaction with Brady, Jensen scrambled out of bed, pulled on a shirt, and then said hello to his new quarterback. Jensen welcomed him to Tampa, congratulated him on signing his new deal, and then after about two minutes of small talk—"Where do you in live in Tampa?" Brady asked his new center—Brady cut to the chase: Could Jensen dry his butt?

The most fundamental aspect of a quarterback's job is to cleanly receive the snap from his center. A natural-born perfectionist, Brady does everything in his power to keep the palms of his hands dry during games, which gives him a better grip on the ball and reduces the possibility of it slipping from his grasp. Brady has long detested it when his center has a sweat-soaked behind, not because he's a germaphobe or thinks it's disgusting, but because it moistens his hands and increases the chances of an errant pass. In New England, Brady had his centers place towels and

40

baby powder on the inside of their pants to absorb the sweat, and now he asked Jensen—as politely as possible—if he'd be willing to do the same.

"We're going to shove a towel down your ass and put powder everywhere," Brady said.

A quarterback had never made such a request before to Jensen, but the center quickly said that he'd basically put anything on his butt if that was what Brady wanted. He thought it was a little odd, but just after a few minutes of speaking with his new quarterback, Jensen believed that Brady surely had to be the most detail-oriented quarterback in the NFL. So for the entire 2020 season, Jensen would be seen running on the practice field and on Sundays with little puffs of white emanating from his backside, small clouds that reminded everyone on the Bucs that Brady was unlike any other quarterback.

‖‖‖‖‖‖‖‖‖‖‖

In mid-April, wanting to throw and run outside, Brady walked into a local public park with a bag of footballs. A city employee stopped Brady, telling him to leave because the park was closed due to a city-wide stay-at-home order. Not long after Brady was escorted away, Licht received a text message from former Buccaneer quarterback Ryan Fitzpatrick, who lived in Tampa Bay during the offseason and had a friend who worked for the city's parks department. *The GOAT gets kicked out of parks*, Fitzpatrick messaged Licht. And so began the avalanche of jokes about Brady trespassing on public property during a pandemic.

Tampa mayor Jane Castor couldn't resist taking a bite out of the apple, announcing the violation in a webcast with St. Petersburg mayor Rick Kriseman during a coronavirus briefing, saying that the GOAT "has been sighted." The mayor's giddy name-dropping created a news story—virtually everything Brady did over the next nine months would be reported on and dissected by the media—and it prompted her to pen a tongue-in-cheek apology on city letterhead, writing, "Tom, my apologies for the miscommunication when you first arrived . . . I couldn't help but have someone investigate the sighting of a GOAT running wild in our beautiful city. No harm—no foul."

Brady's missteps in his new city reminded Arians of another quarterback he once coached: Peyton Manning. Not because Brady had been in the news for what his teammates jokingly called breaking-and-entering and trespassing; rather, it showed Arians the level of Brady's dedication. When Arians was Manning's quarterback coach at Indy in 1998, he went over to Manning's house one evening. On his rookie quarterback's closet door, Arians noticed, were taped directives from his mom that showed what shirts went with what pants and shoes; she had assembled his outfits as if he was still in grade school. It wasn't that Manning didn't know how to put on matching clothes; it was that he didn't want to burn extra minutes thinking about it because it would take away from his time to concentrate on football.

"Peyton kind of struggled with the basics of life back then—he couldn't even use a can opener properly—but it just revealed how focused he was on football," Arians said. "Tom is obviously older than Peyton was back then, but he's a football junkie just like Peyton. The work you have to put in to be great matters so much to Tom. All the best quarterbacks have this trait. Tom was just so desperate to get information and get on the field with his guys once he got to Tampa."

Shortly after Brady was booted from the public park, the headmaster at Tampa's Berkeley Prep—a private K-12 school—received a call from an alum asking if Brady and a few of the Buc players could train on their football field. At first, Joseph Seivold, the headmaster, thought it was a joke, but the alum assured him it wasn't. The school's classes had moved online due to the pandemic, which meant the campus was a ghost-town. Seivold agreed to the request. Within a few days, Brady and a group of offensive players began showing up so early that the school had to turn on the outdoor stadium lights to illuminate the field in the pre-dawn darkness. The workouts on the field with artificial turf continued through June, when Brady and the players had to end their practice sessions earlier than they had hoped because a girls lacrosse camp was starting.

Before the Brady-led practices began at Berkeley Prep, Leftwich told his new quarterback to "learn the guys and have the guys learn where you are going to put the ball, how you're going to see a certain play." Once on the field, Brady communicated what he wanted from his receivers on different pass routes and he discussed the defenses they'd be facing in 2020. Wearing an orange practice jersey and shoulder pads and his new Bucs helmet, Brady—who was playing with a torn left MCL that he had suffered in 2019 but wouldn't have repaired until after the 2020 season—threw to wide receivers running different route combinations, situational drills, and red-zone drills. Most of the player-only practices lasted two hours.

One day Alex Guerrero, Brady's longtime body coach, tagged along. During practice, Guerrero whipped out his phone and opened up a radar-gun app. He measured the velocity of Brady's throws. It confirmed what the Buc receivers and tight ends already knew: Brady still had more than enough juice on his fastball, more than enough power in his arm to hit the twenty-five-yard out-routes outside the numbers that he rarely threw in

New England but would be asked to complete at Tampa under Arians and Leftwich.

During one early session wide receiver Chris Godwin ran a deep route for the first time. Brady unleashed a tight spiral. As the ball whizzed through the warm air, it reached its apex, turned over, and softly nosedived into Godwin's arms. Godwin caught it in stride and with ease. He thought: *What a friendly ball to catch. This is going to be good.*

As the days passed, the temperatures and humidity rose, causing the wide receivers and tight ends to tire. Brady didn't want to overwork his teammates before training camp even started—he especially didn't want them to feel like they were running with concrete shoes due to hot weather—so he often would tell them not to run full routes. Instead, Brady directed them to stand in a certain position on the field and they would work on ball placement, with Brady asking each wide receiver and tight end exactly where they preferred the ball to arrive.

But Brady always remained demanding. He was constantly telling his teammates, *One more time, one more rep.* Most of the players had never been pushed like that during offseason workouts by their quarterback, but no one complained. Six Super Bowl rings commanded respect. "These workouts were so important if we were going to have a successful season," said Clyde Christensen. "We didn't get to have a normal offseason because of the pandemic, so it was really up to Tom to get the guys organized and ready to go. They had to learn each other. Even if you're Tom Brady, you don't just snap two fingers and suddenly have chemistry with the guys you are throwing to. It takes time. It takes work. And no one—not one player—wanted to let Tom down."

During the workouts, Brady constantly was telling the receivers and tight ends about the depth of the routes they should be running and how he wanted them to break out of their routes, explaining in detail his likes and dislikes. In turn, the receivers and tight ends learned how the ball came out of Brady's hand, how it spiraled, where he liked to throw it on different routes, and how catchable his ball was. Defensive backs even showed up, because they wanted to do their part to help Brady—who was still in the early stages of learning the playbook—get ready. "Tom was coaching the players and the players were coaching Tom," Arians said. "The entire team just had a different mindset the second we signed Tom. The standard was set early for the kind of work that was demanded and the expectations were raised for the entire team."

Brady was already exhibiting the same type of work ethic that had been legendary in New England and had long ago won over the Patriots. In New England, he was usually in bed by 8:00 or 8:30 p.m. In the morning, he'd rise around 4:00 a.m. and drive from his home outside of Boston to the facility in Foxborough, usually arriving before 5:30 a.m. Other players would walk into the building at just after 6:00 a.m. and see Brady in the weight room, lifting weights, beads of sweat on his face. "Good afternoon," Brady would say with a sly grin, as if telling that player he was late.

On June 11, Fox 13 in Tampa sent a helicopter to shoot film of Brady and his teammates at Berkeley Prep. The camera eye in the sky revealed that another new addition to the team, Rob Gronkowski, was in attendance. Brady's tight end from New England looked like a new player, a fresh player, a happy player—a far cry from his final days as a Patriot when he labored simply to get down the field.

Shortly after Brady signed his name on the contract, the quarter-back called Gronkowski. The last time the tight end had played was in Super Bowl LIII on February 3, 2019, when the Patriots beat the Rams, 13–3. Gronk made the key play of the game: In the fourth quarter, with the game tied 3–3, Brady launched a ball down the field to Gronkowski, who was double covered while running down the seam. But Gronk dove, snatched the ball, and hung on to it as he hit the turf for a 29-yard gain, setting up the only touchdown of that Super Bowl. He finished with six receptions for 87 yards and was targeted by his buddy Brady seven times, but he also suffered a severe quadriceps bruise during the game. It was one of many injuries he had sustained that season. As he walked off the Mercedes-Benz Stadium field in Atlanta that evening he thought, *I'm done.* He retired two weeks later.

But then Brady phoned his old friend in late March. "Would you come down?" Brady asked.

Gronk replied, "I've been waiting for you to make a move."

Brady told Arians and Licht that Gronkowski wanted to play again, and indirectly suggested that the team should trade for him (the Patriots still owned his rights). The Bucs didn't need a tight end—Tampa Bay had former first-round pick O.J. Howard and the talented Cameron Brate on the roster; the two combined to catch 70 passes for 770 yards in 2019—and Arians questioned whether Gronkowski had the desire to play and was physically fit after spending a year in retirement. "No, he wants to play," Brady told Arians. "And he's in great shape. Trust me."

Arians remained skeptical. While working as a color analyst for CBS in 2018, Arians called one of New England's games. He paid close attention to Gronkowski, noticing that he had braces on his legs and arms and that he struggled to run down the field and separate from defenders. He looked tired. Washed up. Broken. But

Brady was adamant: "I'm telling you his fire is back and his body has healed," the quarterback told his head coach.

Gronkowski was only twenty-nine when he retired in March 2019, but he felt he had to step away because the game had exacted a massive toll on his body. He had recurring nightmares about the hits he'd taken to his head and his legs, his sleep haunted by all the high-speed, violent collisions he had endured every autumn Sunday for nine years in the NFL. Five months after announcing his retirement, the normally effervescent, upbeat, every-sky-is-blue Gronkowski grew emotional when he was speaking at an event in New York, holding back tears as he detailed the physical and emotional pain he had experienced from the quadriceps injury in his final season.

"I want to be clear to my fans," he said. "I needed to recover. I was not in a good place. Football was bringing me down, and I didn't like it. I was losing that joy in life . . . I got done with the [Super Bowl] and I could barely walk. I slept five minutes that night. I couldn't even think. I was in tears in my bed after a Super Bowl victory. It didn't make much sense to me. And then, for four weeks, I couldn't even sleep for more than twenty minutes a night. I was like, 'Damn, this sucks.' It didn't feel good."

But about seven months after uttering these from-the-heart words, Gronkowski told Brady that he was ready to return, that he was feeling as good as he had in years. Licht grilled Brady about Gronk, firing off one question after the next: *Can he run? How is his elbow? Is he in a good place mentally? Does he really have anything left in the tank? Are you positive he wants to play?*

Brady answered "yes" to every query, speaking with such confidence about Gronk's ability that he finally won over Licht.

"We put blind faith in Tom that Rob could still play," Arians said. "It was as simple as that."

In late April 2020, Licht reached out to the Patriots front of-fice, asking New England coach Bill Belichick—who also serves as the team's de facto general manager—and others in the personnel department what it would take to acquire the rights to the tight end. Gronkowski had told Patriots staff members that he would only play for Tampa Bay and that he had no interest in returning to New England, and the two teams cut a deal. In exchange for Gronkowski and a 2020 seventh-round pick, the Bucs sent a 2020 fourth-round selection to New England. On April 21, the trade was made.

Gronkowski couldn't have been happier. He considers Brady one of his closest friends. They've shot commercials together, attended the Kentucky Derby, and spent countless hours in the offseason throwing, catching, and studying. Brady enjoys Gronk's kooky sense of humor and his every-day-is-beautiful disposition. "Everyone wishes in their next life that they can come back as Rob," Brady said. Gronk relishes playing jokes on Brady, such as the time at Fenway Park in 2017 when he snatched Brady's jersey out of his hands before Brady threw out the first pitch. Brady chased Gronk into the outfield, tackling him, both laughing in front of the sold-out crowd like boys in the backyard playing one-on-one football.

Gronkowski also knew the warm Florida weather would be good for his body—Brady had emphasized that to him—and that being close to his mother, Diane Walters, would be good for his mind. Walters lives in nearby Fort Myers, and soon af-ter Gronkowski moved to the Sunshine State, the mother and son started spending time together. The two went on bike rides through the sleepy streets of Fort Myers, shared meals, and Gronkowski often stayed at her house, even though he rented a three-bedroom, 4,500-square-foot condo penthouse in Tampa.

Located on the twenty-eighth floor in the Towers of Channel-side, Gronk's new spread had sweeping views of the Bay, a 1,350-square-foot balcony, and a chef's kitchen.

But it was his mom who really made Gronk—one of five boys—happy. Diana had dedicated so much of her life to her sons, four of whom would go on to make NFL rosters while the fifth (Gordie Jr.) would play professional baseball. When the boys were little, she'd wake at 4:00 a.m. and drive them to various practices in the Buffalo area: hockey, basketball, baseball, football. A gifted cook, she was constantly in the kitchen, using forty-pound boxes of uncooked chicken to make meal after meal. Her specialty was chicken soufflé, which remains Gronkowski's favorite dish, and then the boys would wash down their food with gallon after gallon of milk.

"It's pretty special to be down here and just to go visit my mom an hour and a half away, sleep over, wake up in the morning, and drive to work," Gronkowski said. "The impact on my life she has made has been tremendous and has carried over to this day."

On the field with his new teammates at Berkeley Prep in June, Gronk flashed the suddenness of movement that he had lost in his final season with New England. His weight was down about twenty pounds from his normal 265-pound playing weight, but the connection he had with Brady was as strong as ever. Brady could simply look at Gronk with a certain facial expression and the tight end would know what route his quarterback wanted him to run, how fast he should charge off the line of scrimmage, and the precise angle he should take on his pattern. Gronk also knew exactly where Brady was going to place the ball: high and to the outside if there was a defender close to him; right between his numbers if there was a defender on his back. For years Gronk had excelled at using his wide, six-foot-five frame to block linebackers and defensive backs who tried to cover him and contest his catches, and Brady fully expected to use Gronk this way again in 2020. After a few sessions at Berkeley Prep, it was clear to everyone on the field that Gronk still had the skill and the will to play at an elite level.

Once the NFL allowed teams to completely re-open their team facilities in late June, Gronkowski's first day at One Buc Place did not get off to an auspicious start. Arians was known throughout the building as the commander of the "mask police," as he constantly told everyone in his eyesight to wear face masks to mitigate the spread of COVD-19. Inside the main elevator that carries passengers to the coaches' offices on the second floor, a large sign said in bold letters that only one person was permitted to be in the elevator at a time. But only minutes after stepping into the building Gronkowski ignored that directive. With a team employee standing next to him, Gronkowski rode the elevator from the first floor to the second floor. The doors opened and . . . there stood Arians, who had never met Gronk face-to-face as the head coach of the Buccaneers.

"You guys can't fucking read!" Arians yelled. "That doesn't say 'two.' It says 'one.' What the fuck are you thinking? Read the motherfucking sign! If you fucking disobey the rules around here, you won't be here long."

The team employee froze, his eyes wide, mortified. Gronkowski flashed a goofy, lopsided grin at Arians.

"Nice to meet you, Coach."

And so began Gronk's first official day of the 2020 season in Tampa Bay.

CHAPTER 4

||||||||||||||||||||||||||||||||||||||||||||||||||||||||||||||||||||||||||||||||||||||||||||||||||||

A Training Camp
Unlike Any Other

At the first Bucs staff meeting of the 2020 season, the head coach sounded more like a fist-pounding autocrat issuing edicts to his subjects. No one could go to a mall, a restaurant, a coffee shop, or even to the dry cleaners. Hell, he didn't even want anyone to drive to a gas station unless their car was running on fumes—and if they did desperately need to fill up their tank, they'd better wear gloves before they touched the nozzle. And always—ALWAYS!—everyone had to wear a mask anytime they stepped out of their homes, even if it was just to grab the mail.

In a conference room at One Buc Place in early June, Arians told his coaches that their team was built to win it all this coming season. Their most fearsome opponent wouldn't be anyone on their schedule; it was COVID-19. "We've got to do everything in our power to make sure we don't have any sort of issues with the coronavirus," Arians said to his staff. "This will require enormous sacrifices—more personal sacrifices than you've ever made in your life. I'm doing the same. All I'm going to do is come here to work and then go home. That's it. No one is just going to show up and come into my house. It will be all work. But trust me, it will pay off in the end."

Arians was sixty-eight years old. Assistant coach Tom Moore—Arians's longtime offensive consigliore—was entering his forty-first season in the NFL, at the age of eighty-two. "If any of you make me or Tom Moore sick, I have a gun and I will shoot you in the kneecaps," Arians said. "This is serious shit we're dealing with."

Several of Arians's own family members told the head coach that he should sit out the season. Arians was in the high-risk group for COVID-19. As a cancer survivor, a positive test result could land him in the intensive care unit. But Arians never budged on his position: he was going to lead the Bucs in 2020, no matter the risks. "I'll work and then I'll go straight home," he said to one concerned family member. "That will be my routine every single day. It's going to be a magical year. I've been trying to win a Super Bowl as a head coach for too damn long to miss this chance."

Arians's wife, Chris, instituted a rule: No one could enter their home unless they had tested negative for COVID-19 within forty-eight hours. And even then, the guest needed a damn good reason to come over. "I wasn't messing around," Chris said. "And trust me, everyone knew it." Even Arians's grandchildren in Birmingham, Alabama, knew this rule. They couldn't come to Tampa. Instead, they communicated with Pops and Lolli (short for Lollipop) using a Facebook portal.

At the Bucs facility, Arians oversaw changes that were being made to both the interior of the building and its immediate sur-roundings, including new occupancy limits on the number of peo-ple in meeting rooms, elevators, and the locker rooms and new traffic flow policies. The head coach emphasized to his assistants that they had to set the example for the players; if the staff didn't rigidly abide by these rules, he said, the players never would.

"If we all don't do our part," Arians said, "then we may as well kiss the 2020 season goodbye."

Changes were also made outside of One Buc Place. Before training camp began, a skyscraper-sized banner of Brady—in a number 12 Bucs uniform, about to attempt a pass—was hung on the side of Raymond James Stadium. It was a message to the rest of the league: a new day had dawned in Tampa.

||||||||||||

In July, a time for dreaming and scheming in the NFL, the players were allowed back at the team headquarters at One Buc Place for the first time since the NFL had essentially shut down four months earlier due to the coronavirus pandemic. The majority of the coaches and staff members weren't allowed to interact with players on this July morning—it was a player-only workout, assisted by strength and conditioning coaches—but as soon as the players emerged from the facility and began walking onto the practice field, the coaches and front-office personnel migrated to the second-floor windows that overlooked the field. More than forty on the Bucs staff, each with a mask and six feet from each other, pressed their faces close to the windows, searching for that one player: Brady.

Then, in bright Florida sunshine, wearing his new Tampa number 12 jersey, Brady jogged onto the field, showing no signs that his torn left MCL was bothering him—an occurrence that, improbably, would become a pattern throughout the season. It was a sight to behold, Brady organizing the offense and leading his players in various drills and plays from a playbook that he was still learning. "Holy cow," one coach said. "We really got Tom Brady. Holy cow . . . I mean, this is real. This is actually happening."

Brady unleashed a laser to Gronkowski, causing a "Hell yes!" to echo through the second floor of the building. He feathered a ball between two defenders into the arms of wide receiver Mike Evans. He hit wide receiver Chris Godwin on an out route. "He's still got it," Arians said to no one in particular. "We knew he did and he does. Brother, this is going to be great. Damn, there's a whole lot of swag out on that field."

But it was more than just the throws Brady was making. Even though it was only the first day of practice at One Buc Place, Brady's mere presence made this team already feel very different to Arians and the rest of the coaching staff. When Brady spoke to players, they listened with rapt attention, as if his words thundered down from the Heavens. When he was on the field, there was no goofing around, no wasting time, no simply going through the motions. Practice, for Tom Brady, was serious business. He never got after a player or raised his voice to anyone on Day 1— and rarely would at practices throughout the season—but he calmly talked to his teammates to ensure they were thinking the same things as him on certain plays and understood their assignments. He wanted perfection, expected it even, and he wanted it now. And he demanded the same from himself. When he missed a throw, Brady was quick to point at his own chest, assuring the wide receiver that the misfire was his fault. Brady never lost his cool—his calming, we-got-this demeanor is one of his hallmarks—and promptly moved on to the next play.

From their second-floor perch, Arians and Licht could see that something had changed. No one wanted to disappoint Brady.

That was why an offensive lineman would get so upset when he got beat and his man pressured Brady; why a receiver got so mad at himself if he dropped a pass or ran the wrong route; and why a running back beat himself up if he missed a blocking assignment and a blitzing linebacker forced an untimely throw.

Once it was over, Brady and his teammates walked through the summer heat and humidity back toward the locker room. One practice and Arians knew: he had a quarterback unlike any other he'd coached. This was no small claim: the list of Arians's previous signal callers included Peyton Manning, Ben Roethlisberger, Andrew Luck, and Carson Palmer. They had all enjoyed career-years under the experienced tutelage of Arians, and now Tampa's head coach wondered just how good the Buccaneers could be if Brady could come even close to having a career year at age forty-three, if he could again summon the magic that he had constantly produced for two decades in a New England Patriots uniform, if in his professional winter he could enjoy one more youthful summer.

One thing Arians was sure of was that Brady's first day at the team's headquarters couldn't have gone any better. "Damn, this is going to be fun," Arians said to Licht.

Leaving the office that first night, Arians glanced over at Raymond James Stadium, next to One Buc Place. The stadium lights were off, the structure a hulk in the distant dark. For a moment, Arians allowed himself to dream. The Super Bowl would be played in that very place in less than seven months. With Brady on the team, suddenly the frontiers of what was possible had expanded; anything seemed achievable, attainable. The coach of the losingest franchise in America's four major professional sports thought: *We can win this whole damn thing. I know we can. I know it.*

|||||||||||||

The Buccaneers held their first formal team meeting on July 23, the date that rookies and veterans were allowed by the NFL to return to their team's football facility and interact with coaches and staff. Clyde Christensen, the team's quarterbacks coach, met Brady in the parking lot at One Buc Place. "You've cost me so much money in the playoffs by beating us when I was at Indy," Christensen said, smiling. "I'd probably be retired right now if it wasn't for you."

"Hey, we're going to win a Super Bowl here," Brady said. "We'll get you back all that playoff money."

For the team meeting, Arians addressed his players in the team's 100,000-square-foot indoor practice facility. With the players sitting on folding chairs, spaced out to meet the social distancing standard, Arians grabbed a microphone—the first time in his career he had used a mic to talk to his team. His voice exploded from large speakers. Within an instant, he commanded and captured his audience, pointing toward a stadium that sat a few blocks away. "Right over there is Raymond James Stadium," Arians said. "Somebody is going to be dressing in your locker room for the final game of the season, for the Super Bowl. It may as well be you. We have the talent here to do it. We're going to have to beat COVID. You're going to have to make more sacrifices than you've ever made before. You'll have to be accountable to each other. It will take trust, loyalty, and respect. But we have one ultimate goal and only one: to hoist the Lombardi Trophy."

Arians and the staff were confident that the vast majority of the players were in solid physical shape, even in spite of the pandemic. Anthony Piroli, the head strength and conditioning coach, and his assistant, Maral Javadifar, contacted the sixty-three players on the roster before the April draft to determine how they could work out. Many lived throughout the country in homes with their families. As training facilities and gyms started to shut down in March as a result of the pandemic, Piroli and Javadifar had to be more creative in developing individual workout plans. Space was an issue for many players, forcing the Buc training staff to figure out what the players had access to and where they could go.

Tampa Bay opted to do virtual offseason training. The team used an app that allowed the staff to directly send workouts to players with videos and descriptions of how to perform the workouts. The players could send videos back and ask questions. Piroli and Javadifar were available at all times of the day and night to address any issues they were having.

Javadifar, a former basketball player at Pace University, pretended she was the player watching the video, anticipating what could go wrong or what could be confusing. Piroli and Javadifar recorded every exercise individually and then sent them out to each player. Each video was tailored to each player, which meant Piroli and Javadifar shot hundreds and hundreds of videos. They prioritized running, jumping, and throwing because many of the players lacked equipment. The Tampa strength and conditioning coaches believed the best workout tool the players had was their own bodies, and they designed exercises that would tap into the players' different bioenergetic systems.

If the Bucs had a player in California, for example, the coaches would send him a medicine ball, a speed sled or a prowler, and some bumper plates, which are weights the player can put on the sled. The pieces of workout equipment were so small that players could use them in their backyard or in a field. If a player was confused about anything, they FaceTimed with either Piroli or Javadifar and asked questions. The two East Coast coaches often fielded calls from West Coast players well after midnight.

The Buc staff also emphasized the injury stats compiled from the NFL's lockout season of 2011. The lockout prevented organized, team-led offseason activities, so it was up to the players to make sure their bodies were ready for the season. But many weren't, and that was revealed in the number of players who got hurt that season. There was a sudden increase, for instance, in Achilles injuries immediately after the lockout ended and training camp began. Twelve players across the league ruptured their Achilles in the first twenty-nine days after the lockout. In the previous two years combined, there had only been sixteen total Achilles tendon ruptures.

"We needed them to be in the best possible shape when they got here because we had so much work to do," Arians said. "We were so far behind because we were working in a new quarterback with no OTAs [organized team activities] and no preseason games, which meant we had a hell of a lot of work to do to get him up to speed, even if that guy is Tom Brady. But our guys came in and looked great—at least most of them."

As soon as the team was out on the practice field for the first time, Arians, Leftwich, Licht, and Christensen closely watched Brady, dissecting his every movement, his every interaction with his teammates. The coaching staff estimated that Brady had missed as many as 1,500 snaps with his new teammates because of COVID. He had also lost hundreds and hundreds of hours of in-person film study with his coaches and face-to-face discussions about play designs and concepts. Brady had a massive task before him to be fully prepared for the season. Now the key members of the Bucs organization couldn't wait to see him get started on it.

They couldn't have been more pleased with the player in the number 12 Creamsicle jersey, which signified he wasn't to be touched during drills. Brady made several "wow" throws, displaying the arm strength and accuracy and touch the Tampa staff had seen on tape when they had researched the forty-two-year-old quarterback who would turn forty-three on August 3. One coach even remarked that Brady's incompletions were "good throws," which meant that no one other than the intended target had a chance to catch the ball—a refreshing sight to a team that had been haunted by interceptions the previous season. Brady also was in peak physical shape; he ran through every drill like a player half his age.

Before practice, Brady had pulled out his iPhone and scrolled through several pictures with Christensen that featured Brady throwing the ball. He wanted his quarterback coach to know what his mechanics should look like and to alert him if he noticed the smallest of errors. "If my head starts to tilt outside of my left knee," Brady said, "it means I'm probably tired and this can cause me to miss some throws." Brady also asked his coach to listen to his cadence at the line of scrimmage, making sure he wasn't tipping off the defense of when the ball would be snapped by the tenor of his voice.

On the field, Brady wanted nothing but perfection. If a play didn't feel 100 percent right, even if the pass had been completed, he'd often stop practice for a moment. He didn't want ball placement that was off by even a few inches, or a route that was a step too shallow, or timing that was anything short of flawless. "You're going to run it again and again until you really perfect it with Tom," said tight end Cameron Brate.

After practice, Leftwich, Christensen, Brady, and the other quarterbacks gathered in the main team room on the first floor at

One Buc Place. Rewatching film of the practice, Brady peppered his offensive coordinator and quarterback coach with questions: *How is my footwork on this play? Should I have thrown that ball more to the outside shoulder of the receiver? Was this the right pass-protection call I made? Do my mechanics look off here?* Leftwich and Christensen were astounded by Brady's insatiable appetite for information.

Defensive coordinator Todd Bowles had squared off against Brady eight times when he was the head coach of the New York Jets from 2015 to 2018; Brady and the Patriots beat Bowles in seven of those matchups. "Tom never looked like he wasn't in total control," Bowles said. "He always knew exactly where to go with the ball and then he'd put the ball precisely where the receiver wanted it. And even when you pressured him, he never turned it over. Usually, you can bait a quarterback into making a mistake, but Tom never took the bait. Never. And now that we had him with us in Tampa, we knew our entire team had changed. I mean, he changed everything. You could see that from Day One."

In practice, Bowles watched Brady continually beat his defensive backs in drills. If a corner wasn't playing the right technique or he missed a step on his backpedal, Brady would spot it and sling a completion to his wide receiver just beyond the outstretched hands of the corner. Bowles later called Brady "the most precise quarterback" he'd ever seen.

The other former Patriot at camp was still a work in progress. At first, Arians barely recognized Rob Gronkowski. "Gronk looked eight years younger than he did before he retired," Arians said. "He had lost some weight and he looked great. He looked ready to get going."

Alas, out on the practice field, Gronk struggled to catch his breath after a few drills. "You need to get your ass in Tampa shape, brother," Arians said. "You're not in fucking New England anymore."

"Coach, I feel like I'm running five miles per hour," Gronkowski said, catching his wind.

"Get your ass on the sideline," Arians replied. "Take a breather. Get some water."

Brady still had a long way to go to understand the nuances and complexities of Arians's offense, but the coaching staff expected this—and so did Brady. Arians had already told his new quarterback that it wouldn't be easy for him to run an entirely new offensive scheme, but it was the best option for the team. "Look, Tom, it will be easier for one really smart guy to learn something new rather than having twenty-one other offensive players try to learn a new system," Arians told Brady before he signed with the Bucs.

"I totally get that," Brady replied. "I'm all in. Totally."

In his two decades with the Patriots, Brady operated what was known as the Erhardt-Perkins offense. The creation of this system dated back to the 1970s when Ron Erhardt and Ray Perkins were offensive assistants in New England. The design was tweaked and modernized by Charlie Weis when he joined the Patriots as the offensive coordinator in 2000. The beauty of the scheme was its simplicity: play calls described concepts rather than individual routes. The system also used the same personnel in different ways—a running back could split out to become a wide receiver; a tight end could motion into the slot; and a wide receiver could motion in to line up like a tight end or even a running back. This allowed Brady to step to the line of scrimmage, analyze the formation of the defense, and then attack the defense at its most vulnerable spot without having to worry about substituting in different players.

After Brady signed, Leftwich watched every Patriot offensive play from the previous five years. He was determined to understand the concepts and play designs Brady had run best and then incorporate those elements of the New England offensive scheme into Tampa's, easing the transition for Brady. "I needed to find out what Tom did well and then we as a team needed to figure out what we could do well as an offensive unit with Tom as our quarterback," Leftwich said. "The quarterback *is* the offense. He makes it go. In the NFL, it's all about the quarterback. So I had to do everything in my power to make Tom feel comfortable."

To lower the learning curve for their new quarterback, Arians and Leftwich simplified the verbiage of the plays for Brady. One example: In 2019 a play that called for the quarterback to throw a post-shallow in-route required Jameis Winston to tell his teammates, "62 Z-Post Y Shallow XN." Arians and Leftwich trimmed the language of that play call for Brady to: "62 X-ON." Nearly every play call in the extensive Tampa playbook was similarly truncated.

||||||||||||

Arians had developed his offense with Steelers quarterback Ben Roethlisberger in the summer of 2007. After playing a round of golf near Arians's offseason home in Reynolds Plantation, Georgia, the two sipped cocktails in big comfortable chairs on Arians's porch. As the descending sun bled across the horizon Arians—then the offensive coordinator at Pittsburgh—asked Roethlisberger to help him create a new offense. Roethlisberger was thrilled, and after several weeks the pair had developed the latest incarnation of the "No Risk It, No Biscuit" offense that Arians still runs.

The evolutionary origins of the Arians offense can be traced back to his senior season at Virginia Tech, 1974, when Arians was the starting wishbone quarterback. With blond hair that fell below his shoulders and sporting a thick, wondrous mustache that would have made Jimmy Buffett jealous, Arians looked like the classic 1970s rebel. "And I sure as hell tried to play like one," Arians said.

Virginia Tech head coach Jimmy Sharpe instructed Arians over and over to never play with fear. If, for instance, the Hokies had the ball at their own one-yard line and Arians saw the cornerbacks pressing at the line against the Tech wide receivers, Sharpe wanted Arians to audible out of the called run play to a "Go route" and throw the ball deep. "If there is one word that is not in my football vocabulary it's 'conservative,'" Arians said. "I learned to play that way from Jimmy Sharpe. My whole system goes back to him."

In the Arians scheme, the quarterback has at least two options based on how the defense lines up. One option will give the offense a chance to make a first down; the other option will present the quarterback with the chance to score a touchdown, no matter where the Bucs have the ball on the field. Arians wants his quarterback to have a single thought foremost in his mind: *If I have the right matchup and the opportunity to take a shot at the deep ball, take it.* "I don't care if it's third and three," Arians said, "if our best receiver is in single-coverage and he's running a deep post route, throw him the ball."

Arians expected Brady to struggle with this concept. "Veteran NFL quarterbacks are often hesitant to take a chance and throw the ball down the field," Arians said. "They want to take the easy completion."

When Arians started working with quarterback Carson Palmer in Arizona in 2013—at the time, Palmer was beginning his eleventh season in the NFL—he told Palmer, "Have fun. Throw it deep. This is what we do."

Palmer tilted his head and his mouth dropped, believing that Arians was joking. 'Really, B.A.?' Palmer asked. "I can look at the deep ball like that on virtually every play?"

"Hell yes," Arians replied. He wanted Palmer to look down the field as often as he could. Now he wanted the same out of Brady, even though this offensive philosophy ran counter to Brady's death-by-a-thousand-cuts style that he had mastered in New England, an offense that featured short throws to the sideline to running backs, short throws to his tight ends over the middle, and short throws to his wide receivers on crossing routes.

When printed out, Arians's playbook is as thick as an old-fashioned big-city phone book. It has descriptions and sketches of about three hundred plays with notations about when in a game to use each one. He's built it over the course of thirty years, and in 2020 Leftwich planned to incorporate about a third of the plays in the Arians playbook into any given game plan.

There are nine basic pass patterns in the Arians playbook, which contain the core principles of his aerial attack:

Curl: A long or short pass to a receiver who comes back—curls back—to the quarterback.

Load: A short pass—to the right or left—toward a sideline; often called a pass to the flat.

Pin: A longer pass toward the goalpost to one of two receivers who angle toward the middle of the field.

Divide: A long, deep pass to one of two receivers, each of whom divides or splits the field. The receivers line up on opposite sides of the field and crisscross each other as they run corner routes.

X-Ray: A pass to one of three receivers who each run post routes on three different levels on the field—a forty-yard route, a twenty-yard route, and a five-yard route. A post route, for the record, is when a receiver simply takes off from the line of scrimmage and angles toward the goalpost.

Go: A pass to one, two, or three receivers who sprint straight down the field, which Arians likes to call at least six times a game.

Slant: A quick, short pass to a receiver who runs—slants—toward the middle of the field.

Hitch: A short pass to a receiver who runs five yards downfield and turns back toward the QB to catch the ball.

Jet: A deep pass to one of four receivers who run full speed straight downfield.

These nine routes contain literally hundreds of variations. The curl, for instance, can be run twenty-five different ways. When putting together a game plan, Leftwich—in consultation with Arians—would tweak the routes week to week. He always wanted to make sure that whatever plays his opponent believed the Bucs offense would run and had been preparing to defend on game day, they wouldn't actually see once the opening whistle blew.

On pass plays in the Arians offense, the quarterback operates on a read-rotation system. At the line of scrimmage, Arians and Leftwich wanted Brady to diagnose the defense and, based on what he saw, decide who would be his number one receiving option. If, for example, there were two deep safeties lined up in the middle of the field, he'd know to attack the perimeter of the defense near the sidelines. If the defense lined up heavy on the strong side of the field, then Brady needed to immediately look to the weak side.

Then, as Brady dropped back to pass, Leftwich instructed him to work through his progressions, looking first at the best option, then the second best, and so on—a read rotation. The whole offense is based on receivers beating their defenders in one-on-one situations. If the receiver does that and the quarterback has time to make an accurate throw, Arians and Leftwich believe that every single pass should be a completion.

Arians favorite play is called "80 Go"—a play that in a few months would propel the Buccaneers to the Super Bowl. For "80 Go," the offense employs maximum protection for the quarterback with three receivers running deep, one down each sideline and one down the middle of the field. If all three are covered, the quarterback has the flexibility to check down and throw to a running back flaring out of the backfield. Arians wants a touchdown or a check down with this play. But even this play has a run option for the quarterback: if the coverage isn't favorable for Brady and the offense, then Arians wants Brady to check to a run play.

"It's not just everybody runs a 'Go' route with Bruce," Carson Palmer said. "There's something underneath the route for every possible coverage. Sometimes, based on the defense, you need to get the ball out of your hands quick, and make the defense turn and chase. And Bruce's offense gives the quarterback that flexibility. But the quarterback really has to study and put in the hours to understand where Bruce wants you to go with the ball in every type of situation that the defense will put you in."

As much as he adores the deep passes, Arians wants his offense to stay fifty-fifty in terms of passes and runs in a game. The key is to surprise the defense as to which is which. In 2020 Tampa planned to often line up in obvious running formations—like three tight ends—and throw it; in obvious passing formations, such as three wide receivers split out wide, they plotted to run plays, such as a draw. The goal, says Arians, is to keep the defense "playing off their heels, not the balls of their feet."

After his first practice in Tampa, Brady walked off the field with Christensen. Brady admitted he'd gotten confused at one point and made the improper play call. "It's probably been twenty years since I flat out messed up a call," Brady said. Christensen assured his quarterback that, in time, these issues would go away, telling Brady that he's been coaching Arians's system for six years and he still sometimes used terminology from the playbook from the Colts, where he last coached in 2015.

"It will get better," Christensen said. "You need to unlearn the terminology you've been using for twenty years and continue to learn what we're doing here. You'll get it."

Brady confessed that the transition from one offense to another was difficult. "You're going back a very long time in my career where I really [had] to put in mental energy like [this]," he said. "I have to work at it pretty hard physically still. I put a lot of time and energy into making sure I'm feeling good in order to perform at my best, but mentally I think that's been the thing that's obviously had its challenges. I think you couple that with the coronavirus situation and it became even more difficult."

After one practice, Brady wasn't panicking, but now he knew: the transition from New England to Tampa Bay wasn't going to be easy.

||||||||||||

To help their quarterback learn the Tampa offense, Arians and Leftwich had Brady—in the quiet of the quarterbacks room on the first floor at One Buc Place—put on a virtual-reality head-set called the STRIVR oculus. Here was how it worked: A 360-degree camera was mounted on a tripod and placed next to Brady at practice. The camera saw everything Brady witnessed as he faced the scout-team defense. When Brady placed the headset on and moved his head in any direction, he could see what was happening all around him at any moment in every part of the field. This was particularly useful for identifying blitz packages.

The technology also allowed Brady to watch and evaluate his throwing mechanics, studying his arm angle and release point on his throws. Most quarterbacks are like golfers: if their stroke is "off," even in the most minuscule way, the ball won't hit its intended target. Brady could examine his follow-through, his foot-work, and his balance simultaneously when he unleashed a pass. If he missed a throw in practice, he could quickly pinpoint the reason by examining the play through the eyes of STRIVR from myriad perspectives.

"I've been a fan of virtual reality for two decades," Arians said. "Back in the eighties I told anyone who would listen that if anyone could invent a headset to put on the quarterback that would allow him to visually take mental reps, he'd be a millionaire. But it took until 2014 for it to be readily available. As a quarterback, you see yourself, you see your wide receivers, you see your running back behind you, and you see the multiplicity of defensive schemes that you face. It allows Tom to basically practice at any time without breaking a sweat. It's a wonderful tool."

Throughout camp, Brady could often be seen in an office at the Tampa facility wearing his STRIVR headset. Alone, he saw, he studied, he learned. In these hours, he was slowly—*very* slowly—beginning to understand the concepts, philosophies, and subtleties of an offense that was as foreign to him as having a head coach who called him "brother" and "dude." For Brady, inside the walls of the Tampa facility during camp, the learning never stopped. But time was running out. By the end of the first week of camp, Brady and the Bucs had under 1,100 hours remaining before the season-opening kickoff against the Saints.

|||||||||||||||

Arians allowed Brady to be a coach on and off the field. Brady would tell wide receiver Scotty Miller to pump his arms more on certain routes and he'd calmly explain to Justin Watson, another young wide receiver, that he was too stiff and upright in running his pass patterns and that he needed to work on his hip flexibility and fluidity. Brady was constantly pointing out what players—particularly the inexperienced ones—needed to do better to make themselves better and, therefore, the team more successful. Ninety-nine percent of quarterbacks in their first training camp with a new team would not have the skill or confidence to be able to dispense this kind of feedback and instruction—much less have the players listen to the quarterback and heed his advice. But Brady was different; when he said something—*Hey, shave a few steps off that route*—it was as if it was coming from the team owner, with the players paying devoted attention to their quarterback, hanging on every piece of wisdom that flowed from Brady's lips.

"It pissed me off sometimes because I'd tell a young guy something and he'd just kind of shrug his shoulders and then make the same mistake on the very next play," Arians said. "But when Tom would say the exact same thing that I just said, the player would say 'Yes sir,' and the mistake would be corrected."

Brady even sat in on defensive meetings. Sometimes he'd tell a cornerback that he was tipping what kind of coverage the Bucs were about to employ on a certain play with his body language or how his feet were aligned. He'd tell a safety that he needed to wait a few more heartbeats before inching to the line of scrimmage to better disguise that he was about to blitz once the ball was snapped. The one thing every NFL player wants more than anything is to improve, because it will ultimately keep him in the league longer, which means his earning potential will last longer. And now, even in the early days of camp when he was just getting to know many of his teammates, Brady was doing everything in his power to help his new teammates improve.

"If we ran a bad route or if the ball was incomplete, Tom would say, 'Hey, don't worry about it. We'll get the next one,'" Cameron Brate said. "Then he's really encouraging you when you had a good catch or you ran a good route. But there's been times, too, where he's kind of gotten on guys as well if it's not up to the standard we're hoping for. I think he does a good job of balancing the two."

In some areas, however, Brady's expectations were unyielding. After one play early in camp, the rest of the offense was slow to return to the huddle, and Brady suddenly unleashed a profanity-laced tirade. They had been slow a few times during the team's first scrimmage and Brady had let it slide. But he finally erupted, forcefully telling his new teammates that they were forming bad habits: they needed to hustle back to the huddle and they needed to get on and off the field quicker when substituting. His words were firm and final: *No more motherfucking loafing to the huddle.* "We have to get these things right," Brady said. "We're running out of time." To Brady, small bad habits can eventually grow into bigger bad habits and wind up infecting the entire offense. The issue of loafing was eliminated then and there. Coach Brady had made his point.

On another afternoon during camp, Brady sat in a meeting with the entire offense. When the coaches finished reviewing the practice film from that morning, Brady asked the players to stay in the room as the Buc staff members strolled out. Standing in front of his offensive team, Brady told them that the morning's session had been their worst practice of camp. Brady didn't raise his voice, but his words were flavored with conviction. "We have too much talent for this," he said.

Yet just because Arians essentially allowed Brady to assume coaching responsibilities during practices and meetings, that didn't mean Brady was immune from incurring the full-throated, expletive-laced wrath of his head coach. During one walk-through practice in camp, Brady threw a pass to a wide receiver, violating the Arians rule of no balls in the air during walk-throughs. Seeing this, Mount Arians erupted.

"What the fuck are you doing?" Arians yelled. And thus began a dressing down of Brady, which the entire team watched in silent amazement.

"Tom gets cussed out like everybody else," Arians said. "We don't throw the ball in walk-throughs, and he learned that." Still, by the end of camp, Arians could count on one hand the number of times he had raised his voice to number 12.

||||||||||||

With one of his best friends, Rob Gronkowski, on the roster, Brady seemed more at ease, at peace and happier than he did in New England, where Bill Belichick rarely hesitated to unload his frustration at his quarterback in front of the team if Brady had done something wrong. In New England, Brady was afforded little special treatment. But now? Arians asked for his input on personnel decisions, wanted him to be a coach on the field—especially to the younger players—and told him during training camp that it was OK to take a day off here and there. Special treatment in Tampa? "Damn right I'm going to give Tom Brady special treatment," Arians said. "The dude has done enough and proven enough in the league to earn that."

On August 4, after Brady had reached out to free agent running back LeSean McCoy and asked him to join the Bucs, McCoy signed with Tampa. Hours later, McCoy jogged out onto the practice field, where Brady gave him a big hug before he participated in his first practice with the team. He caught passes from Brady and experienced several pinch-me-if-this-is-real moments as he huddled with the six-time Super Bowl winner. "I have scrimmaged with Tom in the past and we are pretty cool," McCoy said. "But to actually see him work, he is like a general leading the troops."

During camp, the general and Gronkowski filmed a video together that was produced by the team called "Tommy and Gronky." The two sat in beach chairs next to palm trees and put their feet into a kiddie pool, located beyond the practice fields at One Buc Place. In a not-so-subtle dig at the New England weather, Brady said, "It's a little toasty out here today." Brady and his buddy then laughed like they were schoolboys at recess, having the time of their life. The message was clear: they were loving football again.

The host of the video asked a series of questions. "Which is Tom's favorite ring?"

"I know this one," Gronkowski said, his voice rising with excitement. "I know it! Oh wow: Gisele loves this one, too."

Brady shot a stern, you-better-behave look at Gronk, who was fully aware that Gisele had stated several times that she had wanted her husband to retire years ago.

Giggling, Gronk gave an answer that was so truthful he may as well have placed his hand on a Bible: "The next one!"

|||||||||||||

Throughout the season Brady and Gronk enjoyed their time doing the "Tommy and Gronky" show. Sometimes they wrote their own scripts; the Bucs' digital and media team dreamed up other shows. In one episode, a straight-faced Gronkowski asked, "What's it like being a daddy?"

Brady replied, "It's kind of like being around you and my teammates all day."

The pair then told a few dad jokes. Brady said he made one up for his pony-loving daughter Vivi, who was eight.

"Why couldn't the pony sing?" Brady asked.

Gronkowski pondered this question for a few beats, then raised his eyebrows as if a light had turned on in his head. "Ooh! Ooh!" Gronk said excitedly. "Because he was a little hoarse!"

The pair laughed and laughed. The giggles continued well after the show wrapped and the two friends walked back across the practice field to the facility, slapping high-fives and looking like two guys who could hang out all day, swap stories, and never stop chuckling at each other.

Brady's sense of humor stood in stark contrast to the preconceived notions that many of the players and coaches had of him.

At press conferences and on the sidelines—the only places the vast majority of his new team had ever seen him—Brady can come off as distant, evasive, and ultra-intense. "All those years I coached in Indy, I thought that everything that came out of New England was dark. I painted a mental picture of Darth Vader and his crew up there in Foxborough," said Christensen, who was on the Colts' staff from 2002 to 2015. "And then you meet Tom and he's about as nice a guy as you could possibly meet. I had to readjust my thinking. My grandkids and kids had been taught for so long that Tom had been a member of the Evil Empire, so I had un-brainwashing to do."

‖‖‖‖‖‖‖‖‖‖‖

Behind the wheel of a golf cart, Arians zoomed through the Florida sunshine, rolling from drill to drill, position group to position group during a summer afternoon late in training camp. The head coach liked the development of his offense and his quarterback in particular. The entire unit, he believed, was flush with talent, a rocket waiting to launch—Brady just needed the reps and the time on the field to get in sync with the other ten players on the field; then the engines would fire and blast-off would occur.

In 2019 the offensive line had surrendered 2.9 sacks per game—twenty-second in the NFL. But after analyzing the game film, the coaches determined that the sack number was deceiving. "An offensive lineman needs to know that the quarterback is going to be on his spot eight yards behind the center when he drops back into the pocket, and often Jameis wasn't on that spot," said Harold Goodwin, Tampa Bay's assistant head coach and run game coordinator. "He'd leave the place where the offensive lineman thought he'd be and that often produces a sack. A lot of the time the quarterback can be his own worst enemy and he's really the one that causes the sack because he runs into trouble.

"With Tom, he's always right where he's supposed to be, right on that spot. He's not going to abandon his spot because he understands that he could actually be increasing the chances of getting hit if he abandons the pocket. He might slide and shuffle a little, but he's going to stay close to where the play design says he should be. There's a reason he's been in the league so long—he doesn't take many hits. And the reasons for that are that he'll throw the ball away if the play isn't there, he won't move too far from his spot."

After Winston got sacked in games, a concerned offensive lineman often rushed to him, asking him if he was hurt. His response was usually the same: *I'm good, bro. I'm good. Next play.* But Brady was different. When he took a particularly hard hit or sack in his final season with the Patriots, he'd look at the offensive linemen helping him to his feet and say, *This shit hurts, man!* Brady wanted his linemen to take pride—and play with concern and urgency—in protecting him. The one time he wasn't afraid to emphasize that he was oldest player in the league was in those painful moments after being drilled to the ground by a 320-pound, wild-eyed defensive force of nature.

Tampa's offensive line—led by center Ryan Jensen (he of the butt towels), left tackle Donovan Smith, and guard Ali Marpet—had been bolstered in April by the addition of right tackle Tristan Wirfs, who the Bucs selected out of Iowa in the first round of the 2020 draft. To pick Wirfs, Tampa moved up one spot in the first round; Licht and Arians were so sure of his ability to be a high-level Day 1 starter that they didn't want another team to make an aggressive move and trade up to snatch him.

Arians had been burned before. In 2017, when he was the head coach of the Cardinals, Arians had planned to select quarterback Patrick Mahomes with the thirteenth overall pick. But then the Chiefs jumped from selecting twenty-seventh in the first round to tenth by trading their first-round pick, third-round pick, and a future first-round pick to Buffalo. The bold maneuvering allowed Kansas City to draft Mahomes. As soon as the pick was announced, Arians slammed his fist on the table at Arizona's draft headquarters, shouting expletives, shocked that the Chiefs would trade up for Mahomes because they already had a Pro Bowl starting quarterback in Alex Smith. Arians vowed that if he ever liked a player as much as he had liked Mahomes, he wouldn't stand pat. And so with Wirfs he didn't; the Bucs traded a fourth-round selection to the 49ers to leapfrog one pick to ensure they landed their right tackle of the future, a move that pleased the team's new quarterback.

When Brady texted Wirfs after the draft to introduce himself, the rookie thought a friend was pulling a prank on him. But Wirfs quickly ascertained that it was no joke and as soon as the two spoke Wirfs couldn't suppress his awe, telling Brady how much he admired him, how he'd seen him play in so many Super Bowls, sounding more like a starstruck fan than a teammate. During training camp, Wirfs's wonder at being so close to Brady continued. After one practice, Wirfs called his former position coach at Iowa, Tim Polasek, and said, "I'm in the same huddle with Brady. I just want to make him proud." In camp, Wirfs hung on every word that Brady and the coaches said—and it showed. The six-foot-five 320-pound Wirfs, according to the coaches, looked like he had the makings of a future All-Pro, displaying rare athleticism and quickness for a man so large.

Throughout training camp, Arians spent time with every unit of the team—offense, defense, and special teams. He talked to offensive linemen about feet placement on different pass protections. He chatted with defensive linemen about pass rushing techniques. He spoke to his kickers about the psychology of succeeding in pressure situations. He conferred with his assistants, constantly asking them what they were seeing, who was performing well, who was struggling, and who needed to be talked to by their head coach. In Tampa, for the first time in his career, Arians was a true head coach, overseeing his entire team, not just the offense.

This was by design. When Arians interviewed for the Tampa job in January 2019, he strode into the meeting with Buc owners Bryan and Joel Glazer carrying no briefcase, no résumé, no three-ring binder with a three-year plan, and no PowerPoint presentation; the only thing he brought with him was his wallet and keys. Speaking with the confidence of a coach who had twice been named Associated Press NFL Coach of the Year, he said he'd take the position on one condition: that he *not* call the offensive plays. Arians had been a play caller for the majority of his adult life, and it nearly killed him during his final season in Arizona in 2017.

The stress of having two jobs—head coach and offensive coordinator—landed Arians in the hospital multiple times during his career, dating to his first head coaching job at Temple from 1983 to 1988. In recent years he's had cancerous tumors removed from his skin, prostate, and kidney. Beginning in the early 1980s, a few days after the end of every single season Arians would get sick—often *very* sick—due to adrenal failure. In 2017 his wife, Chris, had finally seen enough: she told her husband—her high school sweetheart who she married in 1971—that it was time to retire. If he kept on coaching, Chris thought, one day he would die on the sideline.

"Bruce was not the greatest at delegating authority," Chris said. "Even when he was at Temple, he had to go to the hospital because of chronic migraines. The job just took such a huge toll on him physically. In 2017 we just thought it was time to step away."

Arians agreed and spent 2018 working as a color analyst for CBS. But the travel schedule was brutal: Arians would typically leave his home in Reynolds Plantation, Georgia, at 4:00 a.m. on Thursday. A friend would drive him ninety minutes to the Atlanta airport, where for the first time in his professional life he had to navigate waiting lines, Transportation Security Administration (TSA), and boarding procedures. After arriving at his destination, a car would pick him up at the airport and that evening he'd meet with players and coaches with his broadcast partners Greg Gumbel and Trent Green. On Friday, Arians would attend practices, trying to pick up nuggets that he could share with viewers. And then on Sunday he'd be in the booth with Gumbel and Green.

It only took Arians seventeen minutes of his first regular-season game action to uncork his first dirty word on live national television. When breaking down a run-pass option play in the Week 1 matchup between the Steelers and the Browns, Arians couldn't help himself when he noticed that a receiver wasn't covered. "Wide-ass open," he said as he analyzed the play.

During his season in the booth, Arians took detailed mental notes. For the first time in his career, he saw how other teams practiced, he witnessed the different teaching styles various staffs employed, and he especially paid close attention to head coaches who weren't coordinators, watching how they managed their teams. He felt like he was getting a sneak peek into the lives, methods, and philosophies of the competitors he had been trying to beat for years.

By the end of the season, because of the travel (including a trip to London, where he fell ill and was nearly trapped in his hotel by a Brexit rally), he felt as drained as if he'd coached. "Shit, I may as well get back into coaching rather than do this," Arians told his son Jake. Hearing the news that Tampa Bay had fired Dirk Koetter on December 30, 2018, after the Bucs failed to advance to the playoffs for the eleventh consecutive season, Jake called his father. "Do you want me to reach out to Jason and the Bucs for you?" asked Jake, who is his father's best friend and acts as his agent on certain projects. "They've got a good young team."

"Let me talk to your mom," Arians said. "I'll get back to you."

At the kitchen table in Reynolds Plantation, with the waters of Lake Oconee sparkling in the sunshine outside the window, Chris Arians told her husband that he could look into the Buccaneer job on three conditions: they would rent a house and not buy one in Tampa; she could have an unlimited travel budget to see her grandchildren in Birmingham, Alabama; and her husband would work on his health, which was her gentle way of suggesting he should give up play-calling duties. Arians then phoned his son. "Make the call," he said.

"Are you sure?"

"Yes, I am. Make the call. I'll know if it's right when I'm in the building and talking to the owners."

Jake called Jason Licht, telling him his dad might be interested in returning to the sideline. After a brief conversation with Arians's son, Licht then phoned Arians. When Arians saw it was Licht's number on his caller ID, he didn't even say hello. "Hell yeah, baby," Arians said. A few days later, Arians drove from his lake house in Georgia to Tampa, where he met Licht for dinner in a private room at a restaurant. Arians already had a list of twelve coaches he wanted on his staff—and he believed he could convince every one of them to join him. He had a deep personal and professional connection with every name on his list; he had either coached them as players or been on the same coaching staff with every one of them at some point during his forty-four years in football.

"It's like the heavens have opened up," Arians told Licht. "All the guys I want, I can get because they've either been fired or have been out of coaching. I'm going to get the gang back together."

During the interview the next day with Buccaneer ownership, Arians detailed his vision for the organization, speaking with authority, controlling the room. Tampa had employed a search firm to assist in the hiring process, and at one point during the three-hour interview a member of the firm asked Arians what he thought about the use of analytics in football. "I didn't know my guys would be hitting curve balls on the field," Arians said. "I don't fucking care about what the analytics say when it's fourth and three and what the odds are of making it. Football and decisions like that are about gut and feel and you know I'm always going to go for the kill shot. I'm too fucking old to be conservative. We're going to do everything we can to win and win now."

The Glazers had been warned of Arians's blue language by Licht. The owners also had watched the 2016 Amazon documentary *All or Nothing: A Season with the Cardinals* that featured some classic Arians lines, gems such as "Penthouse now, shithouse in a minute," and "As far as goals go, we have one: putting the fucking ring on our finger." When Arians started dropping F-bombs in the interview, Tampa's decision-makers didn't recoil. Instead, they admired the authenticity of the man sitting in front of them, not to mention the words he uttered about how his entire career had been building to this moment and this opportunity to lead this woebegone franchise out of the darkness and into the light.

"B.A. made it all sound so simple and easy," Licht said. "It was like, 'This is what you do. This is how we'll do it. This is why it will work.' When the owners asked me what my experience with B.A. had been like in Arizona, I shot them straight: I told them he's one of the best coaches, if not *the* best, I'd ever been around."

The owners agreed to Arians's condition that he relinquish play-calling duties. But before they could formally offer him the job, the Glazers asked Arians to take a comprehensive physical. The next morning, Arians walked into St. Joseph's Hospital and underwent a four-hour exam, featuring EKGs, stress tests, and a doctor prodding into places Arians had never been prodded. Once the exam was over, the doctor walked into a private room to deliver the results to Arians.

"Hey, you're not in terrible shape," the doctor said.

"What do you think?" Arians asked.

"I'd give you a C–," he said.

"Dude, I haven't been a C– in five years!" Arians said, elated. "I fucking passed! That's better than I did in most of my classes in college!"

The next day Licht negotiated a contract with Arians's agent—it took less than five minutes to finalize a four-year deal with a team option to extend to a fifth year in 2023. Arians also insisted to the Tampa owners that Licht be given a matching four-year contract with a club option in 2023, which Licht signed a few days after Arians was hired.

Almost before the ink was dry on the contract, Arians called Byron Leftwich, who had just been fired as the interim offensive coordinator by the Arizona Cardinals. Arians wanted him to be his new offensive coordinator in Tampa and, more significant, to call the plays. Leftwich's response: "When do I start?"

||||||||||||

In 2019—Leftwich's first season in Tampa as a play caller—the Bucs finished with the top-ranked passing offense in the NFL and had the league's third best total offense. But now that Leftwich had Brady as his triggerman, he expected the offense to operate at an even higher level in 2020, as long as Brady could learn the playbook and be comfortable directing the Arians system.

Brady was doing his part. Off the field, the Tampa staff could monitor how much time each player spent on his team-issued iPad; Brady consistently ranked in the top three, and was often number one. On the field during training camp, Brady displayed long, sustained moments of grace, the kind Leftwich had never before seen. The offensive coordinator was especially struck by the way Brady's teammates were responding to him. Watching Brady throw one tight, on-the-money spiral after the next in training camp, seeing the players listening to him like he was their professor and they were his students, Leftwich turned to Arians standing nearby and asked. "Have you *ever* seen anything like this?"

Arians smiled devilishly, like a man holding a royal flush at a poker table, then turned away to watch his defense. As tan and fit as he'd been in years, Arians did not look like a man whose doctor gave him a health grade eighteen months earlier of C–. He appeared now to be more in the B to B+ range. And it was no coincidence that during the previous offseason Arians—for the first time in his football life, and in spite of a worldwide pandemic—didn't get sick.

|||||||||||||

On August 28, the Bucs held a simulated game at Raymond James Stadium, the final live practice of training camp. Facing the second-team defense, Brady appeared as comfortable in the offense as he had all camp, a conductor in complete control of his symphony orchestra. His opening drive started at his own two-yard line. Brady's first play was a handoff to Ronald Jones, who gained a few yards. Then Brady unleashed a laser over the middle to wide receiver Chris Godwin for a first down. He then connected with tight end O.J. Howard for six yards, then wide receiver Mike Evans for another first down. The 16-play, 98-yard drive ended with a one-yard touchdown run by Jones.

Brady continued his mastery throughout the afternoon, consistently hooking up with his receivers on timing patterns and deep balls (he threw an arching rainbow to Scotty Miller in the middle of the field for fifty-five yards). Arians, who played the role of head referee, walked over to another coach late in the scrimmage and said, "Tommy's ready."

But was he? The season-opener against the New Orleans Saints was only two weeks away.

CHAPTER 5

Making History— On a Coaching Staff

Sitting in the back of a darkened meeting room several days before the Bucs were set to play the Saints, Arians listened intently to his defensive coordinator. With film of the Saints offense running on a screen, Todd Bowles repeatedly clicked on a button that started and stopped the action, sending alternating flashes of light across the conference room that was filled with defensive coaches and the head coach. Bowles offered a methodical, step-by-step dissection of the Saints offense and detailed the Bucs' plan to slow down New Orleans quarterback Drew Brees. From his back-row chair, Arians didn't utter a word—his faith in his defensive coordinator ran as deep as the waters of the Gulf.

During the season Arians would offer suggestions to Bowles, but he never mandated any changes to the defensive game plan. The head coach would typically walk down the second-floor hall from his office to Bowles's on a Thursday and ask about the upcoming opponent. After Bowles would summarize his plan,

Arians would nod his head and say, "Okay, sounds good." Then he'd stroll back to his office, fully confident that his defensive coordinator would outwit and outcoach the offensive coordinator Tampa Bay would be facing in a few days. "Bruce lets all of us work," Bowles said. "He knows we're going to be aggressive on defense. He lets us play ball."

The relationship between Bowles and Arians stretched back to 1983, the year the thirty-one-year-old Arians became one of the youngest head coaches in college football when he took over the Temple Owls. Bowles was a redshirt sophomore strong safety for Temple, and he quickly was struck by the turn-up-to-ten intensity of the new head coach. One summer afternoon, a few months into Arians's tenure, he ordered his players to meet him at Temple's track. With the heat hovering at 100 degrees, the players ran and ran and ran, with coaches timing the players in 160- and 330-yard sprints. Some players passed out; no player met Arians's standard. The next morning at 5:00 a.m. Arians gathered with his team on the practice field for more sprints.

"Bruce had just come from Alabama and coaching under Bear Bryant," Bowles said. "He was trying to instill Alabama toughness in us. And he did."

It only took a few practices for Arians to notice Bowles's superior football IQ. "He was a coach on the field for us," Arians said. "He absolutely was one of the smartest football players I'd ever met." How smart? In 1985, his senior season, Bowles smashed violently into an offensive player during a goal line drill at practice, dislocating six bones in his left wrist. Arians had entrusted Bowles, a team captain, with more responsibilities than any other defender on the roster: he was singularly charged with making the presnap adjustments for the linebackers and defensive backs. He not only told those players where to line up, but he also voiced on-the-fly calls about whether or not the coverage needed to be changed based on the offensive formation. When Bowles broke his wrist, Arians had to slice his defensive playbook in half, because Bowles's replacement wasn't capable of making the same kind of presnap checks and calls.

After the injury, Arians called his injured defensive captain into his office for a serious talk. Arians knew that Bowles aspired to play in the NFL. "I don't think you're going to be making it in the pros," Arians said. "You might want to start coaching. I think you'd be a natural at it." Arians's initial assessment was way off the mark: Bowles played eight years in the NFL and was the starting free safety on the Super Bowl–winning 1987 Washington Redskins. But Arians was right about Bowles's coaching potential.

A dozen years younger than his former head coach, Bowles began coaching in the NFL in 2000 as the secondary coach for the New York Jets, the first rung on his NFL coaching ladder. When Arians became the head coach of the Cardinals in 2013, he hired Bowles to be his defensive coordinator—a major break for Bowles, who never before had been in charge of an entire defense. "I told the Cardinals that I wouldn't take the job if I couldn't have Todd," Arians said. "I always thought, going all the way back to Temple, that Todd would be a head coach before I was a head coach and I'd be working for him. I needed him on my staff to lead our defense."

Bowles and Arians are polar opposites in their personalities—Bowles seems to never shed his expressionless poker face, even when he's playing a leisurely round of golf or enjoying a casual dinner with friends. But both are as aggressive as any two coaches in the NFL in their play calling. Bowles is a risk-taker on defense, as if he inherited Arians's no-risk, no-biscuit nature like genetic code. He dreams up exotic packages to pressure the quarterback and force mistakes, blitzing his opponents from every possible position. In so many ways, he is the defensive version of Arians.

"As a defensive coordinator, it's important to mirror the philosophy of the head coach," Bowles said. "We are both very aggressive yet thoughtful in how we call plays, but we do it in a different style. Bruce will take a bomb and blow up a defense with one play. He'll also blow you up in his office if you've pissed him off. But I'm like a sniper who quietly approaches you. You never hear me or see me coming. Then my defense will take care of business and I'll be home for dinner and you never even knew I was there."

Through the years Arians and Bowles have been known to talk football for hours at a stretch. Back and forth, they've traded ideas on pass routes, blocking schemes, blitz packages, presnap adjustments, and different pass coverages based on down and distance and field position. They've debated how best to scout, motivate players, and prepare for an opponent. Unlike most coaches, the two don't use a chalkboard; they perceptively visualize complex schemes and communicate them with crystal-clear clarity and recognition—masters of the string theory of football.

So, the day after he was named the twelfth head coach in Tampa history, Arians called Bowles, who had been fired as the head coach of the New York Jets less than two weeks earlier. "You coming?" Arians asked. He then went on and on about the warm Florida weather, the talent on the roster, and told Bowles that they were finally going to win a Super Bowl together. But it wasn't an easy decision for Bowles. Matt Nagy, the head coach of the Chicago Bears, wanted Bowles to be his defensive coordinator as well. Bowles had known the Bears coach since Nagy was a toddler; Bowles played for his dad, Bill—a defensive line coach—at Elizabeth High School in New Jersey. Plus, the Bears had the

NFL's top-ranked defensive unit in 2018, and Nagy wanted to hand Bowles the keys to running this elite defense. It was an attractive offer.

But Bowles turned it down out of loyalty to Arians. "It's hard to say no to someone who has been so good to you for so long," Bowles said. "And it was amazing how so many of us on the staff had been together before. We all knew what to expect, what the plan was, and how we were going to execute it. Bruce didn't have to coach the coaches. We all were on the same page from the very start."

IIIIIIIIIIIIIIII

Turn the clock back to March 2017. Arians was lounging poolside at his home in Reynolds Plantation, Georgia, a Crown on the rocks in hand, the calm waters of Lake Oconee in the sun-dappled distance. Unprompted, as he examined a few of his favorite passing routes in his playbook for the Arizona Cardinals, Arians started riffing to a friend on one of his all-time favorite players, a former NFL quarterback who had been a recent intern for Arians—one who had obviously caught the eye of his boss.

"Mark my words: Byron Leftwich will be an NFL head coach," Arians said back then. "I got to know him when he was a backup quarterback with me when I was the offensive coordinator in Pittsburgh in 2010 and 2011, and Byron has just got a mind for the game. Hell, he can coach my offense better than *I* can coach my offense, and *I* designed the damn thing!"

Arians first spotted Leftwich on television back in 2002, when the quarterback was a junior at Marshall. Facing Akron late that season, Leftwich broke his left tibia in the game. But after a speedy trip to the hospital, he memorably limped back onto the field, the agony carved into his grimacing face. After Leftwich threw a long completion in the fourth quarter, two of his linemen—Steve Sciullo and Steve Perretta—picked up their injured quarterback and carried him down the field. That image remained with Arians two decades later, the coach's textbook example of "grit."

As a backup quarterback with the Steelers in 2010 and 2011, Leftwich occasionally was on the business end of Arians's world-class, F-bomb dropping rants. Arians typically is rough on his backup signal callers—"Bruce will MF the backups to death," said Carson Palmer—because it's his way of sending not-so-subtle messages to his starter. But Leftwich always responded the same way: he patiently listened and then would help tutor the starter, Ben Roethlisberger, in the lesson Arians was trying to teach.

"Byron was always like an extra coach for me when he was a backup," Arians says. "And now he can coach things that I can't in my offense, because he's actually run it from the quarterback position and I haven't. For example, he knows where the quarterback's eyes need to be on certain plays when facing different formations and he knows where the ball needs to go and how fast it needs to go there if we motion a running back out of the backfield and the defense blitzes. As a player he was always so calm, showing the demeanor of a coach on the field. And he's the same way now."

After Leftwich retired in 2012 after ten NFL seasons—he threw for more than 10,000 yards while playing for five different teams—he moved back to his hometown of Washington, D.C., and picked up the game of golf. Swinging the sticks six days a week with a standing 7:47 a.m. tee time, he lowered his handicap to 9. The one afternoon he wouldn't be on different courses in the

D.C.-area was on NFL Sundays. Planted on his living room couch with his toddler son at his side, he'd watch his old coach Bruce Arians stalk the sideline, headphones on, his laminated play sheet in his hand.

Arians was thinking about Leftwich, too. He kept calling him, in fact, offering him a chance to coach with him in Arizona. "I want you to come down and just see it," Arians said. "Come down here just for a couple of days and let me know what you think."

"Bruce and I would sometimes talk after games and discuss different plays and what the thinking was that went into the plays," Leftwich says. "I've always loved the Xs and Os of the game. In elementary school I'd draw up plays during class. In junior high I'd design plays that my coaches let me run. The strategy part of the game is fascinating to me."

In April 2016 Arians's persistence paid off: Leftwich agreed to join the Cardinals as an intern. Each year Arians brought ex-players to work as interns to give them a foot in the coaching door. Arians considers the program a key step to increase the number of minority coaches in the NFL. "I was intrigued enough to give it a chance," Leftwich said. "I had no idea if I'd like it or not."

But it didn't take long for the thrill of football to once again grip Leftwich. As soon as he saw the whiteboards in the Cardinal offensive meeting rooms with play designs, different offensive packages, protection schemes, and various routes, he was hooked. He immediately thought, *This is my world. This is my life.*

During that 2016 training camp Leftwich lived in the shadow of assistant Tom Moore, who had been Peyton Manning's offensive coordinator in Indianapolis from 1998 to 2008. Moore had over sixty years of coaching experience—he remains Arians's most trusted confidant in Tampa—and Leftwich listened hard to every word and each story that flowed from the silver-haired septuagenarian, absorbing his knowledge like nourishment.

In 2017 Arians promoted Leftwich to quarterbacks coach. During his playing days Leftwich, by his own admission, wasn't the type of mobile QB who "could run for ten yards on a third and nine." That meant that, as a player, he needed to understand how to work through his progressions—his throwing options— and how to read a defense in order to attack its most vulnerable area on the field.

This skill served Leftwich well during the 2020 training camp as he worked with Brady. Because Leftwich was an immobile quarterback, he knew exactly what the slow-footed Brady was experiencing as he ran Arians's offense. "Byron knows what it's like to stand back in the pocket and manipulate the defense from the pocket," Arians said. "That's something you have to do to succeed in our offense."

"Tom and I are not what you'd describe as athletic quarterbacks," Leftwich said. "So I knew exactly what it was like for him out on the field. You've got to zip through your progressions— boom, boom, boom—and then get the ball out of your hand, either an on-time throw or throw it away. Tom isn't going to beat you with his legs. It's his brain that beats you. And he's going to do his damage to the defense from the pocket, not outside of it."

IIIIIIIIIIIII

Inside One Buc Place, the staff jokingly refers to itself as "Temple South." Six coaches have a Temple connection. Bowles, running backs coach Todd McNair, special teams coordinator Keith Armstrong, and cornerbacks coach Kevin Ross played there under Arians. Safeties coach Nick Rapone was on Arians's Temple staff. And assistant defensive line coach Lori Locust was a student there when Arians prowled the sidelines. Arians often spoke with the Owls basketball coach John Chaney, who passed away in 2021. Chaney emphasized to a young Arians that he needed to be his authentic self, even if it rubbed people the wrong way. "Don't be afraid to tell the truth, no matter how ugly it comes out," Chaney told Arians. "It's coaching, not criticism."

"Bruce approaches players and coaches the same way at Tampa that he did at Temple," Armstrong said. "So many of us have known each other for most of our lives that it's almost like we can communicate without even speaking. We know the routine. We know the expectations. We know the schedule. And we know that we have to be open and honest with each other. I can't imagine another staff in the NFL that trusts each other as much as we do, and that all goes back to just knowing each other and each other's families and what everyone has gone through in life to reach this point. A lot of head coaches talk about the staff being a big family, but we really are. And the players know it because they see it for themselves, they don't just hear it."

The first branch on the Arians coaching tree is Bowles, the head coach of the Jets from 2015 to 2018. It didn't end well in New York for Bowles: the Jets lost 21 of the last 27 games he coached. Overall, he finished with a 24–40 record in four seasons and failed to make the playoffs. The biggest issue with New York during Bowles's tenure was the lack of a dynamic offense—or any semblance of an offense, actually.

Bowles plowed through three offensive coordinators in four seasons. The organization's biggest mistake of the Bowles era wasn't made by Bowles. In 2016—a year after Bowles led the Jets to a 10–6 record, the last time New York had a winning season—the front office insisted on picking Penn State quarterback

Christian Hackenberg in the second round of the draft. How did Bowles feel about this? As usual, his actions spoke louder than his words: Bowles never allowed Hackenberg to play a single down in a regular season game before Hackenberg was traded to the Raiders on May 22, 2018, for a conditional seventh-round draft pick.

"I had four years to make something happen and it didn't work out. No excuses from me about that," Bowles said. "But I learned a lot of lessons that made me a better coach that I could apply to Tampa."

While he was the head coach of the Jets, Bowles attended offensive meetings, deepening his understanding of what NFL offenses were trying to do to beat defenses—what schemes they were running, what plays they tended to call in different down-and-distance situations, and what went into building a play script for a game. "All of that made me a better defensive coordinator," said Bowles.

Bowles also learned the value of self-scouting. He saw how offensive coaches would zero in on the tendencies of the defenses they were about to face and then develop a plan to exploit those tendencies. This lesson—understanding the value of self-evaluation and coming up with an outside-the-box plan to surprise your opponent—would serve Bowles well at Tampa's most critical juncture in 2020.

||||||||||||

Under Arians in 2020, the Bucs had the most racially and gender diverse coaching staff in NFL history. All three of his coordinators—Leftwich, Bowles, and Keith Armstrong (special teams)—were Black, an NFL first. Harold Goodwin, the assistant head coach and running game coordinator, was also Black. Goodwin began his coaching career as a graduate assistant at Michigan in 1995 when Brady was on the team—"The only thing that has changed about Tom since he was in Ann Arbor is his bank account," Goodwin said—and he served as Arians's offensive coordinator at Arizona from 2013 to 2017. In his first three seasons by Arians's side with the Cardinals, Goodwin helped Arizona to a 34–13 record and two trips to the playoffs, including reaching an NFC championship game. Three teams—the Rams, Bills, and Jaguars—offered him head coaching interviews, but it was clear to Goodwin that those were only "token" requests, made in the spirit of the Rooney Rule, which requires franchises to interview a minority candidate from outside the organization for any coaching vacancy.

At one interview, the owner wasn't present, which told Goodwin he didn't have a legitimate chance at the job. Another team demanded that he keep both of their current coordinators—a bizarre stipulation that few potential head coaches would accept. "I felt like the decision was made about me before I even interviewed," Goodwin said. "I don't really know if the Rooney Rule has any merit. I've been in the league since 2004 and we're still talking about some of the same issues in terms of minority head coaches."

At the start of the 2020 season, there were only three Black NFL head coaches on the thirty-two teams (9.37 percent), in a league where roughly 70 percent of the players are Black. This made Arians's 2020 staff an eye-popping outlier. Arians insisted that the common skin color of his top four lieutenants was a coincidence and that he was simply hiring the most qualified people for their respective jobs, but no one in the NFL who knew Arians well was surprised that he was blazing a new path.

|||||||||||||

Back in 1970, when Arians was playing quarterback at Virginia Tech, an assistant coach named John Devlin handpicked Arians to become the first white player to room with a Black player in school history. Arians didn't think twice about breaking a segregation barrier; his closest friends in his old neighborhood in York, Pennsylvania, were Black. Arians had been the quarterback on his Pee Wee football team—he had snow-white hair and was one of the few Caucasian kids on the roster—and the mothers of the Black players called him "Whitey." He thought it was hilarious.

At Tech, Arians became fast friends with his new roommate James Barber; they hung a sign on their dorm-room door that read "Salt and Pepper Inc." They wore each other's clothes and hit it off as if they'd known each other for years. A few times white players would come by and ask Arians, "What's it like living with a Black guy?" Arians would roll his eyes and shoot back, "I'm sure it's a hell of lot better than living with you. Your shit is dirty all the time and James is the greatest guy in the world."

Arians and Barber became graduate assistants together at Virginia Tech. They grew so close that Arians and his wife babysat Barber's twin boys, Ronde and Tiki—future NFL stars. Tiki was sick a lot as a kid—he had fevers and convulsions—and Arians and Chris would take care of Ronde when the family was at the hospital with Tiki. Arians still fondly recalls bouncing Ronde on his knee for hours.

"Bruce can talk street with anyone and, if he needs to, he can be the most intellectual guy in the room," Bowles said. "Because of Bruce's unique background, he can reach absolutely everyone on a football roster, and that's the key to building chemistry and building a winning team."

Arians also was the only coach in NFL history to have two women on his full-time staff: Lori (Lo) Locust, an assistant defensive line coach; and Maral (M.J.) Javadifar, the assistant strength and conditioning coach. Arians hired Locust and Javadifar before the 2019 season. But not even Arians could have anticipated the depth and breadth of the positive impact that Coach Lo and Coach M.J. would make on the Bucs—and on Arians himself.

|||||||||||||

Arians first noticed the skill of female coaches when he was an assistant coach at Mississippi State in the 1980s. He grew acquainted with Dot Murphy, who at the time coached wide receivers at Hinds Community College in Raymond, Mississippi. Murphy's background was in basketball—she had played two years on the U.S. national team—and from 1977 to 1982 she coached basketball at Mississippi University for Women. During that time her husband was the head football coach at Hinds, and late during one blowout game he let his wife call the offensive plays. The result? The Hinds offense immediately scored a touchdown.

Her husband then hired Murphy onto his staff. In her second year, 1985, Hinds was playing a road game when one of the receivers coached by Murphy dropped a pass. The defensive back looked at Murphy and said, "You don't need to be on the sidelines. You need to get back in the stands." The Hinds players heard the comment. After the game, Murphy had to restrain her players—who were very protective of her—from confronting the mouthy defensive back.

Arians watched Murphy closely and believed she was one of the best wide receiver coaches he'd ever observed. She understood the nuances of the game like a lifelong veteran, which in many ways she was. Murphy, whose father was a high school football coach in Mississippi, grew up playing tackle football with her ten male cousins—she has a crooked right middle finger and chipped tooth to show for it. During practices, Arians saw Murphy make her wide receivers hit the dirt and do ten cutaway push-ups after every dropped pass. He saw her earn the players' respect with an unusual mix of yelling and coddling. Murphy at the time was the only female assistant at any level of college or pro football. In a world of hypermasculinity, she quickly built a trust with her players, which she believed was rooted in the fact that many of them were closer with their mothers than their fathers.

"Dot was terrific, a great teacher, a motivator, and was able to connect with her players," Arians said. "It was clear that a woman could do the job in what up to then had been a man's world. That was when the seed was planted in my head that if I one day became a head coach, I'd like to open a door and give a female a chance to be on my staff if she had the skill and talent. Some women can simply connect to players in ways that male coaches can't."

In 2015, while Arians was the head coach of the Arizona Cardinals, he hired the league's first female training camp coaching intern, Jen Welter. Arians was impressed. The players responded to Welter, who coached linebackers, in ways he'd never seen before. "It was almost like they were trying to please their mothers," Arians said. "It was incredible."

Shortly after being hired by the Buccaneers, Arians came home from the scouting combine in Indianapolis and had a long discussion with his wife, Chris, a lawyer, about bringing on a female assistant. "I think I'm going to hire a female intern," Arians said.

Chris looked her husband squarely in the eyes and shook her head. "You're not going to hire another girl and give her one year and then kick her out," Chris said. "If she's qualified, hire her full-time."

"You're right," Arians replied. The next day he asked Licht and the Tampa ownership if he could hire two full-time female assistant coaches. Everyone gave him the same answer: yes. "The best coaches are teachers," Arians said. "And when I was growing up, most of my best teachers just happened to be female. So I just kept thinking that if I ever found the right female who understood football and was a great teacher, it could really benefit the team. Great coaching isn't just about schemes and what plays you call. Great coaching, to me, is more about connecting with your players and getting them to understand what you are trying to teach them."

In 2018, Locust was an assistant coach for the Birmingham Iron in the Alliance of American Football (AAF). She had been a student at Temple when Arians became the Owls' head coach in 1983 and had met him a handful of times. During her fifteen-year climb through the coaching ranks that weaved across the country and through all levels of the sport, she learned the value of finding creative ways to stick out. So when she found out through a friend that Arians was looking to hire a female assistant coach, she sent Arians an email. The subject line read: "Thirty-six years later, I'd love to work with you." Days later, she was in Arians's office interviewing for a coaching position.

Locust began playing football when she was forty on a woman's semi-pro team in Harrisburg, Pennsylvania. After suffering a knee injury, she started coaching the team and later became an assistant head coach at her alma mater, Susquehanna Township High School in Pennsylvania. For the first year, she mostly just observed, trying to learn as much as she could about coaching. She also attended coaching clinics around the country, passing out business cards and hoping for a chance at a higher level. At the clinics she never saw another female face. Whenever she'd have to use the restroom, she'd usually have to go on extended walks; the women's rooms at the clinics were typically converted into men's rooms, because it was rare for a woman other than herself to show up at any coaching clinic in America.

In the fall of 2018 she interned with the Baltimore Ravens and then became an assistant for the Birmingham Iron of the short-lived AAF. She convinced Iron coach Tim Lewis to hire her by telling him, "Coach, I just want you to know that you don't have to make up any accommodations for me. I've been doing this for a long time. I've been coaching men's football for a long time. You can say, do whatever you want. I'm good."

"Lori was just one of the guys," Lewis said. "No one held anything back on what they said or did. Football transcended everything else. Everyone saw early on that Lo could do the job. She knew her stuff and she connected with a few players that none of the other coaches could. Her unique skill set made that happen. She helped our entire team become a family, which is the goal every season, and she helped build our team chemistry."

In Tampa, Locust found her voice—and won over the players. "In our first meeting with the defensive line we went around the room and everyone stood up, introduced themselves, and told the room a little about themselves," said Kacy Rodgers, Tampa's defensive line coach. "Coach Lo related all the obstacles she had overcome to make it to the NFL and the guys could really relate to her. From that moment forward, she was just Coach Lo to all of us. It was like we didn't even notice her gender. Her overall football knowledge was impressive, but she had never coached in the NFL before and the NFL is a completely different game than any other level of football. Coach Lo was a sponge. She took just as many notes as the players in our meetings when we were going over installs. If I said it, she wrote it down."

One of Locust's first assignments in 2019 was to teach the playbook to recently signed defensive tackle Ndamukong Suh. Before they even talked, Locust researched Suh's background. She knew that on the field he was as violent as anyone in the league—in 2011 and 2012 Suh had been voted by his peers as the NFL's dirtiest player in polls conducted by the *Sporting News*. But after speaking with coaches who had worked with him, she also discovered that he had a softer side—an emotional side—and that he was very intelligent.

They started reviewing the playbook together. The playbook was relatively new to Locust, so she spent hours at night in her bed learning every nuance of the Bucs' defense—the formations, the play calls, the responsibilities of every player in every scheme they ran. Then Suh and Locust sat down and started studying together. She was the teacher; he was the pupil. Over a short time, they established a connection and trust. Suh could see that Locust knew her stuff.

Then they got out onto the field. One day the Bucs defense was performing a drill that required the defensive linemen to run forward five yards after the initial contact with the offensive lineman. Midway through the drill Suh started loafing, only running one yard. "That shit is fucking ridiculous!" Locust yelled. Suh and the rest of the players looked at Locust, almost stunned. But she had to make everyone realize that she was the coach, she was in control. Because Suh and Locust had already built a relationship, he took to her coaching. Suh lined right back up and did the drill, correctly.

"I never would have yelled at a player like that if I hadn't already solidified and nurtured a relationship," Locust said. "Relationships are key to coaching. You can't push every player the same way. You must know your limits with everyone. The only way to do this is to understand the boundaries with each player. And the only way to know those boundaries is to really know each player—what motivates them, what they are playing for, what could trigger their temper to flare, and what the best way to reach them is. As a woman, I've always had good intuition, and I rely on that every day at work. But that intuition is grounded in the facts I acquire about each player as I develop professional relationships with them."

||||||||||||

Her background intrigued Arians: Maral (M.J.) Javadifar played basketball at Pace University in Westchester County, New York, and then earned her doctor of physical therapy degree from New York Medical College. She worked as a physical therapist and performance trainer for the sports teams at the University of Virginia and in 2018 landed a job as a physical therapist at Avant Physical Therapy in Seattle. Arians strongly believes in "sports science," and he wanted a coach with a background in both physical therapy and training who could be a bridge between the training staff and strength coaches. After he hired her, Arians charged Javadifar with a unique responsibility: she would work with the players to help prevent injuries before they happen.

"The players love M.J.," Arians said. "She is relentless and she pushes the players as hard as any coach on our staff. Hell, she pushes me when we work out together—and even kind of scares me sometimes."

A few weeks after taking the job with the Bucs, Javadifar was working with a player who had a problem with his upper thigh. As he lay on a training table, she assessed him, asking him to describe the pain he was feeling. "You know, Coach," the player said, "it's kind of like when your balls are tingling. You get what I mean?"

Not knowing what to say, she deadpanned, "Actually, I don't know what you mean." The player, horrified, looked at her with eyes that were as big as saucers.

Seconds later, they cut up and high-fived at the absurdity of the situation. But in reality, this was a breakthrough moment for Javadifar and this player. He had forgotten about her gender; she was just his coach and trainer. They never talked about his "balls" again, but it certainly made them closer, this unspoken secret joke they shared. Soon they were giving each other knowing head nods in the locker room—the ultimate sign of acceptance in the workplace.

When Javadifar started working with a player for the first time, she tried to read his body language and pick up on his nonverbal cues. "If his hands are folded across his chest or if he's looking away or if he appears to be sucking on his teeth, then I'll know he's hesitant and not sold on me," Javadifar said. "It's so important to pick up on these cues. You need to know the mindset of the man you're trying to help before you really try to work with him. You need to understand what you're about to deal with."

If she sensed a player had doubts about her ability to coach and guide them, she'd become very matter of fact, ask him about his injury and his goal. Then she'd actively listen to the player and would tell him, "I'll show you how I can make a difference." Then she'd do a movement assessment. During the 2020 training camp she worked with a cornerback who was having trouble with his back. At first, he wouldn't look her in the eyes, which told her he thought she had no business being in the training room. Then she checked his standing rotation and measured it. After that, they did some stretching and an exercise routine. Through it all, he wouldn't make eye contact.

Once they were done, she again measured his standing rotation. It increased by several inches. Now, suddenly, his eyes started meeting hers when she spoke. The player didn't say anything, but it was clear that he was beginning to believe in Javadifar's methods. "Results matter and sometimes you have to be patient with a player to get those results," Javadifar said. "And when you achieve a positive outcome, the question of gender disappears. Players want one thing: to get better."

During training camp in 2020, Locust and Javadifar were typically at their desks by 4:30 a.m. They'd open their laptops and review their schedules for the day. They'd often come up with three plans of action/reaction for everything that could happen in the coming fifteen hours—plans A, B, and C. "You always need to expect that things won't go as you anticipate, so it is vital to have well-formulated, well-thought-out backup plans," Locust said. "Preparation is critical for us, because any misstep we make could be construed as an example of us not belonging here."

At 5:30 a.m. the players started coming into the football complex. Javadifar would assess the players in the training room, examining how their bodies were moving and trying to get to the root of any physical dysfunction they were experiencing. Both coaches liked to joke with the players—they could drop MFs with the best of them—and they always tried to make the players feel at ease.

"Female coaches in the NFL should be more of the norm, because they bring a different perspective to the job," Arians said. "We need female coaches because they will make players better in ways that men can't. They can connect to players in ways men can't, they can empathize in ways men can't, and they often can listen in ways that men can't. I'm telling you, Lo and M.J. are just the beginning of what's going to be a revolution in NFL coaching circles. More women are coming."

During the camp, Javadifar not only worked with the players, but she also helped Arians. On many mornings he trekked to the weight room, where she would stretch the head coach and direct him through various physical therapy exercises to minimize the pain he was experiencing in his back and legs. After several sessions, with the start of the season drawing closer, Arians felt as physically good as he had in years. "M.J. is the best," Arians said. "Even coaches wear down. And just like players, if you feel good as a coach, you can do your job better."

|||||||||||||||

At one of the final practices of the 2020 training camp, Locust jogged from player to player on the practice field, her intense eyes squinting in the sun, critiquing defensive linemen on their hand placement during a drill. Amid the grunts and the cracks of colliding helmets—the soundtrack of every NFL practice—she blew a whistle and yelled, "Go!"

At the same time, inside the facility in the training room, Javadifar reviewed the rehab schedule with a linebacker trying to overcome a knee injury. Then she walked over to another player, helping him stretch his legs. As the player winced in agony, she smiled and offered a few encouraging words.

Arms folded, Arians slowly walked across the practice field like a philosopher deep in thought. He had the coaching staff he wanted. He had the quarterback he wanted. And now at this moment, here on this sun-kissed afternoon, his wildest dreams stretched farther than ever. At this moment, he genuinely believed he was presiding over the best team of his coaching life.

CHAPTER 6

Brady's Struggle—
On and Off the Field

They sat in a second-floor conference room at One Buc Place, the quarterback and the offensive coordinator, reviewing the game plan one more time. It was Friday, September 11, and the season-opener against the New Orleans Saints at the Mercedes-Benz Superdome was less than forty-eight hours away. The Bucs had just finished their final run-through practice of what the offense and defense were expecting to face in New Orleans; Brady and the offense spent extra time on short-yard situations and goal-line plays with blitzes. Now Brady and Leftwich were meticulously fashioning together the Bucs' opening series of plays for the Saints game.

Coaches have been scripting the initial plays for games since Bill Walsh, the former head coach of the San Francisco 49ers, pioneered the practice in the late 1970s. Walsh's opening script consisted of fifteen plays. Tampa's sheet has thirty. In these plays Leftwich—as taught by Arians—involves as many skill position players as possible. He wants the running backs to have a few carries and he tries to give all of his wideouts a chance to make a catch or two, wanting his skill position players to take an early hit and get acquainted with the game. The logic is that the earlier your best players become engaged in the action, the better they'll perform down the stretch. That's why Leftwich and Arians view the first thirty offensive plays as absolutely vital to the team's chances of success.

In the conference room, with a whiteboard in front of them that listed all 150 potential plays the Bucs might run against the Saints, Leftwich selected the first 15 running plays that he was going to call. Then Brady named the 15 pass plays he wanted on the script, a mixture of drop-backs and play-action passes. In New England, Brady didn't have as much input into the game planning—he told his boyhood idol Joe Montana that Patriot coaches would ask for his opinions and then often ignore them, which led to a "beef" between Brady and the New England staff—but now Arians and Leftwich wanted to hear anything and everything their starting quarterback thought about the opening script and overall game plan.

Arians and Leftwich never will call a pass play that their quarterback doesn't believe will work. After more discussion, Brady then circled his four favorite plays on the board, which would be the first throws he would make against the Saints. The meeting lasted about an hour. Leftwich and Brady would meet again on Saturday night at the team hotel to put the finishing touches on the opening script and overall offensive game plan—a routine they would continue throughout the season.

"Tom looked great in our final scrimmage," Arians said. "He was in command and the throws were crisp and on time. We told him if the play wasn't there just to throw it away. Don't force anything and don't take any unnecessary hits. He was ready to go—at least I thought he was."

|||||||||||||

The silence was surreal. When the Buc players jogged onto the Louisiana Superdome field for pregame warm-ups on September 13, they weren't greeted by boos or jeers or any cheers—the only sounds they could hear were the voices of other coaches and players and their own footfalls on the turf. Due to COVID-19 restrictions, the Superdome's 73,000 seats were entirely devoid of fans. To Brady and other Tampa players, this immediately cast upon the season-opener—typically one of the most overheated, fever-pitched environments of the season—the feel of a glorified scrimmage.

Near midfield, with the players loosening up, Arians greeted Jameis Winston, who had signed with the Saints to back up starter Drew Brees after being released by the Buccaneers. Arians gave his former quarterback a hug and asked him about his wife, Breion. The Winstons had just gotten married, and had a second child on the way as well. "Congratulations on the wedding and the baby," Arians said. "Family is everything. I'm so happy things are working out for you."

"Thanks, Coach, and thanks for everything you did for me," Winston said. "The GOAT is going to be great for you." This was a generous moment for both the coach and the opposing quarterback, cementing what Arians already knew: Winston harbored no hard feelings toward anyone in the Tampa organization.

Before the game, as they would always do, Arians and Leftwich closely monitored Brady. They wanted to determine if he looked tense or uptight; if the ball was coming out of his hand nicely; if his drop backs were hurried; and whether he was in rhythm. Arians always preached to his quarterback to stay in the moment—"The precious present," as basketball coach Rick Pitino describes it—and not look too far ahead. In these pregame inspections, Arians examines the body language of the quarterback. "The rest of the offense feeds off the quarterback," Arians said. "So if your quarterback looks shaky before a game, you need to do something about it, and damn fast." Arians has used a multitude of tricks through the years to get the mind of his quarterback in the right place if he detects something is bothering him before kickoff, including talking about his family, telling him he needs to focus on his footwork (and therefore quit worrying about the opponent), or asking him if he has any good memories of playing at this particular stadium.

But Brady appeared as smooth and as at ease and as on time as ever with his throws in the pregame. His mechanics were spot on, his footwork flawless. To Arians and Leftwich, it was a beautiful sight.

||||||||||||

Tom Brady has never had the strongest arm. And he's always known it. Once in New England, about a decade ago, he sat with Belichick in the quarterbacks' room at One Patriot Place watching film. As they looked at tape of Jets quarterback Mark Sanchez, Belichick noticed a receiver who had just broken free from the cornerback trying to cover him—about seventy-five yards away from where Sanchez was running for his life from pass rushers. Belichick said aloud, "Just throw it. You're not going to get any more open than that."

Brady was flabbergasted. He could never heave the ball that

far, and he knew it. Brady's reaction revealed a larger truth about him as a player: He's always been acutely aware of his own physical shortcomings. When he drops back for a pass, he doesn't view the unfolding action through the lens of endless possibility; he sees it through the prism of how to maximize what he's capable of doing for the betterment of the team, even if that means making a four-yard check-down flare pass to his running back or even just holding on to the ball and taking a sack. Each week, he studies opponents and pinpoints their weaknesses—all starting NFL quarterbacks do—but Brady always does against the backdrop of his deep-seated self-knowledge predicated on: *How can I be proficient and efficient against this defense given my own deficiencies as a player?*

This is not to say that Brady's liabilities are sizable—his Super Bowl rings prove that—but it shows a self-awareness and self-understanding that is unique at the NFL position. Winston, his predecessor in Tampa, has a stronger arm than Brady, but he couldn't master the art of comprehending his own limitations. He tried to zip balls past defenders (and often failed). He tried to squeeze passes into tight windows (and often failed). And he tried to make throws that simply no human being could complete, which often ended up in the mitts of opposing defensive backs.

Now that Brady was behind center, Arians had a quarterback who understood that the single most important aspect of playing the position is to take care of the football and not give it to the other team. Still, the lessons that Belichick emphasized at every practice—always live to play another down, minimize risk—were now about to be applied to the Arians offense that featured more difficult deep throws (typically at least six a game) and thus more risk than any other offense in the NFL.

|||||||||||||

As Brady warmed up in the Superdome, Christensen thought about the second act of Peyton Manning's career, when Manning left the Colts and led Denver to a 24–10 victory over the Carolina Panthers in Super Bowl 50—Manning's final game. Brady had mentioned to Christensen several times the allure of duplicating what Manning achieved, and Christensen told him that he now had what Manning did in Denver: a pair of elite wide receivers.

"Peyton put up some big numbers in his final years in Denver with [wide receivers] Demaryius Thomas and Emmanuel Sanders," Christensen said. "And I told Tom, 'We have a real fine 1 and 1A in Mike Evans and Chris Godwin. You can do exactly what Peyton did. You have the weapons.'"

Brady expressed to his new quarterback coach that he wanted to test himself to see if he could surmount the same challenges Manning faced when moving to a new team in a new town with a new coach, new teammates, and a new playbook. "I want to try and do that," Brady said to Christensen. "I want to do what Peyton did. The weapons here are amazing."

It was true: Brady now had his best pass-catching corps since 2007, when Brady guided the Patriots to a perfect 16–0 regular season record by throwing for 4,806 yards to the likes of Randy Moss, Wes Welker, Donté Stallworth, Jabar Gaffney, and Ben Watson. Even though New England had gone 12–4 in 2019, his final regular season with the Patriots, the team lacked elite talent at the wide receiver and tight end positions. Brady threw for only 4,057 yards in 2019, his lowest output when playing 16 games since 2010, causing whispers to emanate across the league that Brady's arm talent had significantly diminished. Manning heard that same talk in his final season, but still won a Super Bowl at age forty-one, which at the time made him the oldest quarterback to capture the Lombardi Trophy—and the only one in history to win it with two different teams. Now Brady was older than Manning, another spark that lit his internal furnace and put it all full blast as the opening kickoff against the Saints approached.

|||||||||||||

The start of the season against New Orleans couldn't have unfolded any better for the Buccaneers. After forcing the Saints to punt on their first possession, the Bucs took over at their own fifteen-yard line. Brady went to work. Using the opening script he had devised with Leftwich, the Bucs methodically marched down the field. On their third offensive play, Brady lofted a pretty deep ball to Godwin on the right sideline for twenty-nine yards. He found running back Ronald Jones in the left flat for eight yards. He threw in the direction of Evans, which resulted in a pass interference call and a twenty-two-yard gain. Then, from the two, Brady tucked the ball into his gut and followed a mass of human anatomy—the backsides of his offensive linemen—into the end zone. His teammates surrounding him, Brady emphatically spiked the ball. One drive, one touchdown—just as Brady and Leftwich had drawn it up.

"We were off and running," Arians said. "That was the Tom we had seen in training camp. None of us were surprised."

But then, as quick as a cat pounces on a mouse, the Saints defense stiffened and effectively shut down Brady and the Buc offense for the remaining fifty-two minutes of the game. Early in the second quarter, with the score tied at 7 and facing a second and nine from their own 26, Brady threw a deep pass down the left side of the field for Mike Evans. The Saints were in "quarters" coverage, meaning New Orleans had four defensive backs lined up across the back end of the secondary, each covering a quarter of the field. The deep middle of the field was open, but Evans cut his route off short and ran to the left side. The miscommunication between the quarterback and the receiver led to an overthrow by Brady and the pass was intercepted. It only took the Saints three plays to drive 35 yards, concluding with a 6-yard touchdown run by Alvin Kamara. New Orleans led, 14–7.

Early in the third quarter, Brady made another mistake, one that ultimately doomed the Bucs. Trailing 17–7, Brady dropped back to pass and spotted wide receiver Justin Watson, who was running a speed out route to the right sideline. Brady fired, but was off the mark. The ball needed to be delivered high and to the outside of Watson, but instead it was low and to the inside of his target. Saints cornerback Janoris Jenkins jumped the side-line route, cradled the ball into his arms, and ran untouched for 36 yards into the end zone. The play recalled all the struggles that Jameis Winston had endured in 2019—as well as Brady's final throw of 2019 that was returned for a touchdown in the playoffs against the Tennessee Titans—and it gave the Saints a 24–7 lead. New Orleans ultimately won the game, 34–23.

In the locker room, Brady found Christensen. "Hey, I know what I've got to do to fix this," he said, his voice thick with confidence. "We'll start working on this immediately. But I know what I've got to do."

After the game, Arians went down a path that Belichick never traveled: he publicly criticized his quarterback. He said that, on his second interception, Brady can't throw the ball "low and in-side" on the out route. "He was a little bit late on it," Arians said. "And probably a better decision to go somewhere else with the ball . . . He knows how to bounce back. He knows he didn't play very well. It's not what he expects of himself, nor do we expect. I would anticipate him to have a little more grit, a little more determination this week."

Arians's truth telling sent social media atwitter; pundits across the country wondered if the Arians-Brady marriage was already on the rocks. On his Sirius radio show, Brett Favre said, "The last person you want to call out after the first game of the year is Tom Brady. Dissension could easily enter quickly." But Arians and Brady privately joked about Arians's comments becoming low-hanging fruit for the talking heads on ESPN and other national outlets to chew on.

"Can't disagree with you, B.A., that I didn't play a good game," Brady said. "We'll figure it out."

"It's only one round of a sixteen-round match, Tom," Arians said. "Just one game. We'll get it figured out. Trust me."

"I know we will," Brady said. "I know we will."

∣∣∣∣∣∣∣∣∣∣∣∣

The first-game offensive struggles were not a shock to Christensen. "If this had been a typical year, this game would have been our first live scrimmage," Christensen said. "Tom looked like a stranger running our offense. We were a sixteen-cylinder car hitting on eight cylinders. It was ugly. On some plays, Tom didn't even know where his receivers were or where they were going. It's tough to play well under those circumstances."

A day after the game, Arians was in his office when he saw a commentator on television question why Brady had thrown so many out routes against the Saints—a route he didn't often throw in New England. Yelling at the screen and with no one else in the room, Arians said, "You didn't watch him in training camp, motherfucker! I did. Shut your fucking mouth!"

||||||||||||

September was a difficult time for Brady, especially in his personal life. That month his parents in California visited an urgent care center after both were coughing and not feeling well. An hour later, Tom Brady Sr., seventy-six, found out that he tested positive for COVID-19. The next day his breathing became more labored. He went to the hospital and would spend eighteen days there. He developed pneumonia.

"It was life or death," Brady Sr. said. "They didn't know if I'd make it or not." Concerned and wanting to connect with his father, Brady would FaceTime with his namesake during his drives to and from One Buc Place. His mother, Galynn, who underwent chemotherapy treatments for breast cancer during the 2016 season, also tested positive for the coronavirus, but she experienced milder symptoms and was never hospitalized.

During Brady Sr.'s hospitalization, he was so weak that he struggled to lift his head, found it difficult to focus on anything for more than two or three minutes, and could barely speak. Struggling to stay alive and lying on his stomach, he had 100 percent oxygen pumped into him. He was so sick that he didn't watch the Saints game—the first game of his son's high school, college, and pro career that he had ever missed. As his condition worsened, Brady Sr. heard the nurses mention the word "ventilator," which raised the family's level of concern even more.

Brady called the hospital every day. Hearing his son's voice, Brady Sr. could tell how stressed he was. Not only was he trying to turn around the fortunes of an entire franchise, but now his thoughts were constantly focused on his father, some three thousand miles away, gasping for breath. His father worried about his son, but not as much as the son fretted about his dad. Brady Sr. was Tom's hero—always had been.

|||||||||||||

My, how Tom loved his parents and his entire family. The youngest of four kids to Tom and Galynn, Tom Brady Jr. was the Brady's only boy. He loved growing up in San Mateo—located twenty miles south of San Francisco—on a quiet street with three older sisters, mainly because he didn't have to share a bathroom with them and he didn't inherit any hand-me-down clothes. "They were pretty easy on me," Tom said of his sisters. "They'd bring all their girlfriends over to the house. It was pretty cool."

Growing up in a devout Catholic family—Tom was an altar boy at his local parish—young Tommy learned the value of hard work at a young age. He landed his first job delivering newspapers before he was even a teenager, his mom often driving him through the streets of San Mateo in the early-morning darkness with Tom hurling newspapers onto doorsteps from the passenger seat of the car. Even then, he had a pretty good throwing motion.

Tom played every possible sport as a kid except one—football. He excelled on the soccer field, the baseball field, the basketball court, and on the golf course, but he didn't start playing organized football until he was in high school. Arians always emphasizes that he believes young athletes should be well-rounded and not concentrate solely on football until they are in high school. "I want you to be a great bowler and be great at badminton," Arians said, "because down the road that will help more at becoming a great quarterback than merely playing the position and only that position as a kid."

Tom was driven to succeed at an early age, pulling out the same you-don't-believe-in-me card that he would use repeatedly as an NFL player. But as a kid, this was directed precisely at three people: his older sisters, who were all bigger, faster, and stronger than him when he was a boy. "They were the best athletes in my house," Tom said. "They were certainly better athletes than me. They all had successful athletic careers. It was a very athletic family. We grew up on baseball and soccer fields. I loved just tagging along. I loved going out there and cheering them on. I was living and dying with every win and loss that they had."

By the time he was a senior at Junipero Serra High School, an all-boys Catholic school in San Mateo, it looked like Tom was going to be a professional athlete—in baseball. A catcher, he threw right-handed but batted left-handed. He had a powerful arm—he could gun down opposing would-be base-stealers with pinpoint throws, displaying a gift that would one day carry him to legendary status in the NFL—and he hit .311. In one game, he blasted a ball so far over the fence at an opposing stadium that it hit the rooftop of the team bus, startling the napping bus driver, a story that his teammates still tell to this day. During his senior year he was invited by the Montreal Expos to take batting practice at Candlestick Park, where the Expos were playing the San Francisco Giants. Tom again belted a few balls over the fence during BP, prompting one scout to later gush that Tom "could have been one of the greatest catchers ever."

But by then Tom was obsessed with football. He played linebacker on the Junipero freshman team; he was still too raw to play quarterback. They didn't win a game all season, finishing with eight losses and one tie. Tom learned what it was like to be on a losing team, knowledge he turned into more motivation. Even then, he understood the importance of practice, of taking the game seriously, of listening to his coaches. The summer after his freshman year he obsessively worked on what was called the "five-dot drill," where he would put down in his yard five cones—"Dots"—in the shape of the dots on a die and then he'd run various patterns, touching each cone. It didn't matter if his parents had company over or if it was before the sun rose, Tom could be seen in his yard running through the five-dot drill at all hours. Over the span of a few months, the once clumsy kid suddenly looked like one of the most coordinated athletes in all of San Mateo.

Tom became the starting quarterback on the Serra High junior varsity team as a sophomore. In his team's first game of the season, Serra was down by five points with less than two minutes to play in the fourth quarter. Tom calmly talked to his teammates in the huddle—remember, most of these players had never won a high school game before—and reassured them that he was going to lead them to victory as long as everyone did their job and remained focused. With less than thirty seconds to play, Tom threw the winning touchdown pass, prompting an oh-my-god celebration on the field and in the stands that would prove to be a harbinger of Brady's future. It affirmed for him, for the first time in his life, that there are no shortcuts in football. Everything about the game could be distilled into two words: hard work.

Brady went on to start for the varsity team his last two years of high school, throwing for 3,514 yards and 33 touchdowns over that span. His dad, his hero, never missed a game, watching his son with pride from the metal bleachers. After the final whistle blew—win or lose—father and son would always talk, analyzing what had just happened, assessing what Tommy could have done better, rehashing what he'd done well, and just enjoying being together. According to a former teammate, they were the best of friends.

Tom never wanted to disappoint his dad. If he attended a high school party, drank alcohol, and woke up with a hangover—which he did on a few occasions—he'd feel guilty in the morning. Why? Because he didn't want to do anything in secret that his dad wouldn't condone. His tight relationship with his father kept Brady on a road that was paved with right decisions.

As the boy Tom Sr. still calls Tommy aged, the bond between father and son remained as strong as ever. A few days after the Patriots won their first Super Bowl championship in 2002 over the Rams in New Orleans, Tom Sr. flew to Boston to spend some time with his son. Before Tom Sr. returned home to California, his son gave him his Super Bowl ring. "Dad, it's yours," he said. But Tom Sr. said he couldn't take it, no matter how touching the gesture, because he hadn't earned it. He gave it back. But when Tom Sr. arrived back in California, the son called to tell his father to check his carry-on suitcase. Inside, tucked in there by the son when his dad wasn't looking, was the ring.

||||||||||||

Early in his career, Brady could usually catch a few hours of fitful sleep after a loss. But when the Patriots lost to the Giants, 17–14, in Super Bowl XLII—Brady's first defeat in the big game—he stayed awake all night, replaying each possession in his head, snap by agonizing snap. He thought about the throws he missed, the reads he didn't make, the presnap adjustments he should have called, the sacks he could have avoided—no mistake avoided strict scrutiny.

Now Brady usually doesn't sleep after any loss. He didn't get much rest after the Patriots were defeated in the 2019 regular season finale against the lowly Miami Dolphins, costing New England a first-round bye in the playoffs, and he didn't slip into dreamland after the Patriots were eliminated from the playoffs the following week against the Titans. Brady's former Michigan coach Lloyd Carr once said that he'd never seen a player enjoy the chess match and the struggle of football more than Brady, but when Brady can't solve the riddles that the defense presents him with on the chessboard, it haunts his sleep, causing him to twist and turn the night away.

Even when a quarterback has full command of the offense, it's still the hardest position to play in sports. About twenty-five seconds elapse between the moment the quarterback walks to the line of scrimmage and scans the defense to the whistle when the play is over. Dozens of decisions need to be made by the quarterback in those twenty-five seconds: *Do I change the play based on how the defense is lined up? If so, what should I change it to? If the play is a pass, what receiver will be my hot receiver if there's a blitz? Is my offensive line in the right protection? Does my running back know where to pick up the potential blitzing linebacker? Are the defensive backs playing zone or man coverage? Are the safeties creeping toward the line of scrimmage or are they hanging back? Where will the hole be in the defense that I can attack? Where are the strongest spots in the defense that I need to avoid? Is there a weakness in the defense? Where is it? Can I exploit it? How?*

Then there are the four seconds between the snap and the whistle. "The called play in an NFL game only works about half the time," Arians said. "When things don't go as planned is when the quarterback really has to know the offense. And Tom has to process so much information in such a short amount of time, just like any quarterback. Twenty years ago NFL defenses typically had ten different coverage formations and five different blitz packages. But now a quarterback will see that many in the first two series of a game."

Another change in the game compared to two decades ago: back then the same eleven defenders essentially stayed on the field for the entire game, unless they became injured. But now defenses will use up to twenty different players in a game (the Saints used eighteen defensive players in the season opener), many of them specialists whose singular, ultimate job is to make every play miserable for the quarterback. This makes it infinitely more difficult for the quarterback to know at the snap of the ball which receiver will have an advantage and which defensive lineman is most likely to win his matchup against a certain offensive lineman in front of him. No other position in sports requires this much on-the-fly thinking, and it is a reason why Brady is so tormented by losses, because, in defeat, the failure is as much intellectual as physical.

Brady thought he needed to do a better job of understanding his protection—he was sacked three times against the Saints and pressured on several throws. The defense can always blitz one more player than the offense can block, so at the line Brady was determined to improve his reads—his assessments—of what he thought the defense was going to do, and then identify where his hot receiver was going to be if a safety, a linebacker, or multiple defenders came on a blitz. If he had time to throw, then Brady would go through his proper progressions—one, two, three, four, five. If the safeties split to the edges, Brady's read progression is from inside out. If the safeties rotate inside to the middle of the field, his progression is just the opposite, outside in.

Against the Saints, Brady missed several of what Arians calls "rhythm throws"; when Brady's back foot hits the end of his drop—*whoosh!*—the ball immediately flies out of his hand. The rhythm throw is one of the hardest passes in football, because the ball must leave the quarterback's hand before his receiver makes the final cut on his route. This takes timing and practice—and *more* timing and *more* practice. "We didn't have enough reps yet before the Saints game, simple as that," Christensen said.

Brady also misfired on a few "hitch throws" in the Superdome: when Brady reaches the end of his drop, he hitches up a step, sets his feet and—*whoosh!*—he flings the ball. "Timing is everything in the NFL," Arians said. "An NFL quarterback either has this skill or he doesn't. I'll take a quarterback with a great sense of timing any day over one with a big arm who struggles to make rhythm and hitch throws. Our timing was just a few beats off against the Saints."

|||||||||||||

Several hours after the loss to New Orleans, Brady lay awake in the master bedroom of the Jeter mansion, reliving the game, analyzing what went wrong, what he needed to do better, why he threw those two critical interceptions. Throughout the night, the questions floated through his mind like ghosts: *Can we play at a high level? How long will it take to fully understand this offense? What do I need to do better in leading my teammates?*

Even before the sun broke over the horizon, Brady was behind the wheel of his black Ford Raptor, headlights on, driving through the predawn darkness to One Buc Place. Monday was evaluation day for the Buccaneers. The entire team would be in the facility watching the Saints game with their position coaches, reviewing in frame-by-frame detail what went right, what went wrong, and how every player could improve. Brady was the first player in the building.

He knew how much work he had to do.

‖‖

CHAPTER 7

A Gem Washes Ashore in Tampa

Tom Brady had been here before. Three times in his career Brady had lost season openers: 2003, 2014, and 2017. After each defeat, he then led the Patriots to a win the following week. And in each of those seasons, he guided New England to the Super Bowl.

Brady projected an image of calmness—even borderline serenity—all week during meetings and practices. There was an economy of motion in all of his actions, as if he was saving his energy for the moment he trotted onto the Raymond James Stadium field on September 20 to face the Carolina Panthers in Tampa's home opener. On the practice field, he moved with a graceful, elegant ease, never expending more effort than was required, never raising his voice, never showing a hint of panic or confusion. "The players saw Tom's confidence," Arians said. "They knew his track record in bouncing back after losing season openers. This is why Tom is so special: just his presence influences the other players."

"I've never been around a player as calm as Tom," Leftwich said. "The guy is never rattled. And when other players see that, they calm down. Never seen anything like it in my life."

And Brady felt a responsibility to his teammates. He didn't want to let them down. He was still cultivating and nurturing new relationships, forging the bonds that are so critical to building overall team chemistry and creating a winning football culture that extends from the locker room to meeting rooms to the practice fields to Raymond James Stadium on gameday. So at home at night, alone in the Jeter estate, his iPad playbook aglow, Brady continued to study the Arians offense. Leftwich was still streamlining the verbiage of the play calls, trying to simplify what Brady had to communicate to his players in the huddle and make the calls as brief as possible. But Brady knew the reality: it was up to him and him alone to understand every term and concept in this new language. It was a challenge he had never before faced in his career, and even though he had looked very much like a quarterback in the final, futile days of his career against New Orleans, his belief in himself remained superhero strong. He did what he always did when faced with a problem in his athletic life: he worked to solve it, spending more hours during the week before the Carolina game studying the playbook—as measured by the Buc staff—than anyone else on the team.

Las Vegas didn't show the same belief in Brady. After their Week 1 loss, the Bucs were listed as a +1600 longshot in many sports books to win the Super Bowl.

|||||||||||||

Gamedays early in the 2020 season just felt weird to Arians, with no fans or family allowed to attend the game because of the pandemic. During his career he usually had friends and family stay at his house on Saturday nights. Everyone would rise early, eat a big breakfast, share football war stories, and then Arians would shower and head to the stadium. But now, on the morning of September 20, it was only himself and his wife of forty-nine years at the breakfast table.

Bruce and Chris married in 1971 at St. Rose of Lima Church on Market Street in York, Pennsylvania, and they still fondly recall their wedding story. Their reception was held at Tremont restaurant, a cozy place where Chris's grandmother once cooked meatloaf and potpies for diners. Later that night they hopped into Chris's ten-year-old Buick Special and drove halfway to Blacksburg, Virginia, where Arians played at Virginia Tech. They spent the night at a little roadside hotel in Staunton, Virginia—their honeymoon.

Now Arians drove alone in his Mercedes 550 convertible to Raymond James Stadium, wearing his Kangol hat, waving to fans along the way, arriving three hours before kickoff. He walked into his private office, where he sat for thirty minutes, contemplating the offensive and defensive game plans, considering tweaks and adjustments. He was confident that Brady would play at the high level he had displayed in practice all week.

He then strolled into the locker room, where music was blaring—Arians always lets his players pick their own pregame music, believing it keeps them relaxed—and the players appeared loose. Arians looked into the eyes of nearly every player, gauging whether or not he appeared ready and focused. The coach liked what he saw. He patted a few backs and then wandered out onto the sun-drenched field, where as usual he paid close attention to Brady, making sure his tempo and balance and throwing motion were textbook perfect. They were.

As the home-opener kickoff against the Panthers neared, Arians gathered with his team in the locker room. He then raised his hand, signaling he was about to speak. Never one who has believed in the power of soaring, lyrical pregame speeches, Arians typically delivers a business-like, plainspoken message to his players; he doesn't want to raise the emotional temperature of his players too high before kickoff. Now he told his team in a steady, calm voice, "Let's play the best thirty minutes of football we've ever played. Then at halftime we'll make all the adjustments we need to make. We got this. Play smart, play fast, and play with passion."

Arians then led the mass of his players through the darkened corridors of Raymond James Stadium and into the tunnel that leads to the field. The players waited at its mouth while Arians strolled out near the fifty-yard line. As cannons belched fire, the players charged out onto the field—a typical pregame routine, but now there were no fans. Arians looked up to the coach's box, where Chris sat in 2019. He always blew her a kiss at this pregame moment—a touching tradition they had maintained across three decades—but now the Arians box was empty.

The national anthem played, and Arians's eyes grew misty. Even after hearing its notes at hundreds of pregame ceremonies, he still experienced a full-body shiver of emotion when "the rockets red glare" echoed through the stadium.

Arians was ready.

||||||||||||

It was a simple, straightforward play, one that kids could draw in the dirt. With 1:46 remaining in the fourth quarter against the Panthers, the Bucs held a 24–17 lead. On first and ten from

Carolina's forty-six-yard line, in the silence of an empty stadium, Brady called his cadence at the line of scrimmage—everyone on the Panther sideline could clearly hear Brady's voice. He received the snap from center Ryan Jensen, took three steps backward, and handed the ball to Leonard Fournette.

Fournette had been on the team for less than two weeks, and in practice Arians and Brady noticed something different about Fournette: they could *hear* him run almost as much as they could see him. When he made contact with a defender, it wasn't a thud; it was a loud crack. And when he ran, the coach and the quarterback heard his feet pound the ground, a rhythmic, piston-driving sound at full speed.

Now against the Panthers, Fournette powered straight ahead into his line. He blasted between the tackles, busted through the arms of two defensive linemen, broke free from a scrum of players, juked a defender, and then cut up field in full thunder. He sprinted into the secondary, flashing his 4.5 40 speed. A safety had the angle on Fournette, but Tampa's newest player outran him to the end zone, sealing the Bucs' 31–17 win.

Brady sprinted after Fournette, though he hadn't gotten any faster since his 5.28 at the combine all those years ago. He was the third Buc player to reach the running back, and he wrapped his arms around him and slapped him on the helmet. "This is why you're here!" Brady said as they jogged back toward the Tampa sideline. "This is the reason!"

Fournette smiled wide, his teeth gleaming like piano keys. At this moment he may have been the happiest professional football player in America.

||||||||||||

It was an ordinary workday for Leonard Fournette. On August 31, 2020, the running back parked his car outside of TIAA Bank Field in Jacksonville, Florida. He walked through the early-morning shadows of the stadium, heading toward the players and coaches entrance to the Jaguars football headquarters. Before reaching the gate, though, a team employee stopped Fournette, who Jacksonville had selected with the fourth overall pick in the NFL draft three years earlier.

"Coach wants to talk to you," the employee told Fournette.

Confused, Fournette was led into the office of head coach Doug Marrone, who told his starting running back to take a seat.

Freeze the moment here, with Fournette sliding into a chair. In 2019, Fournette had authored the best year of his career, rushing for 1,152 yards on 265 carries, catching 76 passes for another 522 yards, and tallying just one fumble in 341 touches. When Tom Coughlin, who, in 2017, was the team's executive vice president of football operations, picked Fournette over the likes of Patrick Mahomes and Deshaun Watson, he believed that Fournette would be Jacksonville's long-term feature back in his 1980s style, grind-it-out, run-first, pass-second offense. "He's special," said Coughlin on the day Fournette was selected. "We need playmakers. We need people to put the ball into the end zone."

Fournette had been labeled as *special* since he first started playing football at New Orleans's Goretti Playground. He was so big and powerful at age eight that opposing youth league football coaches demanded to see his birth certificate. As an eighth-grader he played on the freshman team at St. Augustine High in Slidell, located just northeast of New Orleans where Leonard's family moved when he was ten. Early in his freshman season, his coach inserted Fournette—the team's starting running back—as a nose tackle. He finished the year with eleven sacks.

When he ran onto the field for his first varsity game as a freshman at St. Augustine, the legend of Leonard already was in full flower in the area. In practice, eighteen-year-old seniors were often hesitant to tackle him, fearful that the fifteen-year-old freshman carrying the ball would injure them. Before he'd even played in his fourth high school game, LSU offered him a scholarship—the first time the Tigers had ever extended an offer to a player that young. As a high school senior, Fournette was the top-ranked recruit in the nation. LSU coach Les Miles compared him to Michael Jordan, and no one in the Bayou State thought Miles was exaggerating.

In high school he was treated like the star he was. The city's police superintendent became his personal trainer. A team of professionals—a masseuse, a chiropractor, and a doctor—all worked on his body before and after games. A filmmaker tailed him, tracking his every move for a documentary.

He lived up to the hype at LSU. In his first season in Baton Rouge, he set the school freshman rushing record with 1,034 yards. The next year he broke school records for rushing yards in a single season (1,953) and touchdowns (22). Following his junior season, which was interrupted by a severe ankle sprain, Fournette became the first player in college football history to announce that he would be sitting out his team's bowl game (the Citrus Bowl) to prepare for the upcoming NFL draft—an act that has now become a trend.

Many NFL scouts viewed Fournette as the second coming of Adrian Peterson; the most famous football player in the city of New Orleans, Archie Manning, said Fournette reminded him of Jim Brown. At the NFL combine, the six-foot-one, 240-pound Fournette ran a 4.51 40—the fastest 40 time for a player that heavy in fifteen years.

Tom Coughlin was sold. But after rushing for 1,040 yards as a rookie, Fournette began to fall out of favor with the organization's decision makers in 2018. That season he got into a fistfight with Buffalo linebacker Shaq Lawson, which resulted in a one-game suspension. Fournette threatened to fight a fan in Tennessee. Then he missed the final game with a foot injury. Watching the action from the bench, he appeared disinterested, unwilling to support his teammates on the field. Coughlin lit into Fournette afterward, calling him "disrespectful" and "selfish."

Following the 2019 season, the Jags declined Fournette's fifth-year option on his rookie contract. New general manager Dave Caldwell and head coach Doug Marrone tried to trade Fournette. He didn't fit into the plans of new offensive coordinator Jay Gruden, who ran a West Coast offense that would feature multiple running backs, not an every-down, feed-me-the-ball workhorse like Fournette. Plus, the team didn't know how he'd react when the number of his carriers was reduced. But Caldwell and Marrone couldn't find a trade partner. "We couldn't get anything, a fifth, a sixth, anything," Marrone said.

So now on this August 2020 morning in Marrone's office, the onetime legend of Louisiana, the former can't-miss running back who some believed was destined for the Hall of Fame, was sitting in front of his coach. Without offering a detailed explanation of why, Marrone delivered the news: *We're releasing you.*

Fournette left the building in a daze, wondering what had just happened. For the first time in his professional life, his own team didn't want him. He drove home and played with his three kids, taking them to a park and riding bikes. He contemplated retiring from football. He ended up clearing waivers—no team put a claim on him—and this deepened his sense of confusion. He was free to sign with any team. *No one wants me*, he thought. *What is happening?*

The following day he received a text. Encouraged by Arians, Brady sent a short message to Fournette. *Man, I would love for you to come help us out*, Brady wrote. Reading those words, something happened to Fournette, something that welled deep inside of him and rose like a flower in a springtime garden. *I want to play again*, he thought. *I want to play with Tom.*

The Buccaneers were looking to add depth to the running back position. Ronald Jones was the starter and the team had recently signed veteran LeSean McCoy, but Arians and Licht wanted another back on the roster who could carry a heavy load in case Jones suffered an injury. Soon after Fournette cleared waivers, Arians walked into Licht's office.

"What do you think about Fournette?" Arians asked.

"Let's look into this," Licht said. "There could be something here, but let's make some calls and find out more about him."

Arians went back to his office. He watched every carry of Fournette's from 2019—analyzing everything from his hip flexibility to his demeanor in the huddle—and Arians liked what he saw. Fournette still possessed a rare combination of power, speed, and agility. But Arians needed to dig deeper. Would Fournette be willing to accept a diminished role in Tampa, where the staff had already decided to use multiple players in the backfield?

Arians picked up his cell phone to call Terry Robiskie, who was on the Cleveland Browns coaching staff with Arians in 2003 and had been Jacksonville's running back coach in 2019. One of Arians's secret weapons is the contact list in his cell phone. It contains the numbers of hundreds of coaches with whom he has worked through the decades. His network of former colleagues and friends extends to every level of football—high school, college, the pros—and every corner of the country, from New York to Los Angeles, from Dallas to the Dakotas.

"If there is a football coach in America who knows more coaches than Bruce, I'd like to see him," Licht said. "I remember one time Bruce and I were at Lucky Dill, a restaurant in Tampa. I was talking to the bartender for a few minutes and then went back to the table and Bruce was on his phone. I asked him what he was doing. 'Oh, I'm just texting with this coach from William & Mary about this guard.' He said it so casually, but it just reinforced to me that if he needs any information on a player, he can usually go straight to the team that player is on or has been on because it's very likely Bruce will know somebody on that team's staff. It's been a vital asset for us."

As Arians called Robiskie, he sat in his office a few feet from a print of Paul "Bear" Bryant. Wearing his customary houndstooth hat, Bryant is holding a piece of chalk in the picture and is about to write on a board. This image is sacred to Arians: the DNA of Bryant still flows through his coaching blood. And the influence of Bryant was one reason he was now contemplating signing Fournette.

|||||||||||||||

It was January 1981 when twenty-eight-year-old Bruce Arians drove his little Pontiac Astro into Tuscaloosa, Alabama, to talk to the Bear for the first time. Arians had spent the previous three years as an assistant at Mississippi State, where, as the passing game coordinator, he had been a part of the Bulldogs' 6–3 upset of Alabama the previous fall.

Bryant was looking for a running backs coach. Arians's former coach at Virginia Tech, Jimmy Sharpe, had set up the interview—the first time Arians realized how valuable contacts and having people vouch for you are in football. Arians had learned from Sharpe that Bryant had been impressed with his work with Dave Marler, a former kicker that Arians taught to play quarterback at Mississippi State. Marler had thrown for 429 yards two years earlier against Alabama.

"Go visit with coach for ten, fifteen minutes and see what happens," Sharpe told Arians.

And so he did. Arians walked into Bryant's office and took a seat on a little sofa, which he swears had the legs cut off. When Arians sat on the sofa, he had to look up at Bryant, giving the Bear a home-field advantage in his office.

Bryant sat behind his massive oak desk, a string of smoke rising from the red-gray ash at the end of the unfiltered Chesterfield dangling from his lips, silently inspecting Arians for a few moments. Then the Bear said in his gravelly, pack-a-day drawl, "I hear you have a way with young Black players. Is that true?"

Bryant knew that Arians had grown up in a predominately African American neighborhood in Pennsylvania and that many of his closest friends were Black. "Well, Coach," Arians said, "I don't know about that. I don't care what color the kids are. Hell, they can be green, red, white, or gold for all I care. But I do know that I'm going cuss them out if they screw up."

"I don't allow cussing," Bryant said. "It's a dollar a swear word."

"Shit, looks like I won't be getting a paycheck," Arians replied.

Arians thought he had blown the interview, but Bryant liked this brash young coach and hired him to oversee his running backs. Beginning on Day 1 in Tuscaloosa, Arians watched Bryant closely, viewing the head coach as more a kindly teacher than a grizzled boss. Bryant knew the name of everyone in the building—from the janitors to the cafeteria workers to the secretaries—and was familiar with their backstories. If a coach's secretary was having a bad day, the Bear would know about it and stop by her desk to offer a few encouraging words. "It was magic the way he dealt with people," Arians said. "He could read others better than anyone."

The Bear, who sometimes played his three top quarterbacks in a game, treated each one like family. Arians watched as Bryant, each Saturday morning before a game, would take a slow walk with his top QBs outside the team hotel. As they strolled together, Bryant would tell them how much he believed in them, how much he cared for them, and how proud he was of each one. Bryant's stern, stark stare could make you cry for your momma, but Arians also saw a softer side of Bryant, a compassionate side, a loving side. Why did Bryant's players work so hard for him? Arians believed it was because they knew he loved them—a lesson Arians would never forget.

Bryant normally watched practice from his famous tower overlooking the field. Arians's running backs stretched right under the tower, and if they heard his footsteps, it usually meant the Bear was coming down to unleash holy hell on a coach or a player. One day Arians thought he was about to incur the full-throated wrath of Bryant. Before practice Arians had been reviewing film with his players when he heard a knock on the door. It was Miss Linda, Bryant's secretary, and she wanted to know if Arians's running backs were going to practice that day. "Yes, we are," Arians said. He then looked at his watch: practice started in minutes.

Arians and his players ran out of the meeting room. The players quickly dressed in the locker room and hustled onto the practice field. They didn't get a chance to stretch; the horn blew to signal the start of practice as Arians and his running backs bolted out of the locker room.

The players knew they were in trouble. They knew Arians was in trouble. But then, under the intense guidance of Arians, the players proceeded to have the best practice of the season, because all of them were on edge—including their coach. After practice was over, Arians and his backs stayed on the field to go through a few extra drills. In one blocking exercise, Arians ran at the players holding a large pad; his players flipped him over their heads. As Bryant started to descend the thirty-three swirling steps on his metal tower, most of the coaches had already gone inside to the locker room, but some of the older coaches stayed around to witness the approaching certain ass chewing.

Then Bryant emerged from the tower and rolled up to Arians and his players in a golf cart. The players couldn't wait to see their coach get his tail whooped.

Bryant walked up to Arians. "Shit, ya'll ought to be late more often," he said. "That was the best damn practice y'all had all year." Then the Bear drove off. It was another lesson learned for Arians: a head coach who is a little bit feared is a good thing.

Arians's final conversation with Bryant took place only weeks before Bryant passed away on January 26, 1983. Arians, who had been on Bryant's final staff, had recently been named the head coach at Temple, and he returned to Tuscaloosa to say his goodbyes. He walked into Bryant's office and the two embraced. Arians asked him questions about recruiting and organization, and then Bryant told Arians about the nature of coaching—and life, really. "You get a job," he said, "and you do a hell of job at it. You look for a better job until there ain't no better job. Then you work your ass off to keep that job."

Then, before Arians left his office, Bryant told him to carry one piece of wisdom with him for the rest of his days.

"Coach them hard," Bryant said, "and hug them harder later."

Those were the last words Bryant ever uttered to Arians. They became Bruce's guiding philosophy.

IIIIIIIIIIII

After a bit of chitchat with Robiskie, Arians began quizzing his former colleague about Fournette, asking about his character, his practice habits, his physical ability, and whether he could accept playing for a team on which he wouldn't be the featured, every-down back. Arians has a well-earned, time-tested reputation around the league for his straight talk and his inability to BS when it comes to football matters; in turn, coaches don't spin the truth to Arians when he asks them direct questions. Robiskie didn't. He gave Fournette a glowing recommendation, even though his bosses had just decided to cut him. Arians thanked Robiskie for his time and told him to stay in touch.

Arians went back to Licht's office. They decided they were going to give Fournette a shot.

Licht called Fournette's agent and it was clear from the start of the conversation that Fournette wanted to sign with Tampa Bay, even though a few other teams were interested. Less than forty-eight hours later, on September 2, Fournette put his signature on a one-year contract with the Bucs that had a maximum value of $3.5 million.

Before the deal was announced, Licht ducked into the quarter-backs' office room, where Brady was studying film. "Bruce and I are going to sign Fournette," Licht said.

"Oh, hell yeah!" Brady replied. "Let me know if you need me to help with anything."

With his text, Brady already had.

||||||||||||||

Now Brady and Fournette stood close together on the Tampa sideline, awash in sunlight, watching the final seconds of the fourth quarter against the Panthers. Throughout the game, Brady and his receivers were more in sync than they had been in the season opener in New Orleans—a result of extra work in practice on timing routes. Brady finished the game 25 of 35 for 217 yards and one touchdown. Wide receiver Mike Evans, who had only one reception against the Saints and whose miscommunication with Brady had led to an interception, caught seven passes for 104 yards and a touchdown. And Fournette came off the bench and rushed for 103 yards on 12 carriers. The offense still wasn't humming at full throttle, but it improved enough in Week 2 for the Bucs to level their record at 1–1 with a 31–17 victory over the Panthers.

"We are still a long ways from where we need to be," Brady said. "Consistency and dependability are going to be the things we really need. We've got to get back to work. The clock is ticking."

In the locker room, Fournette was still aglow as he sat on a chair in front of his locker—one teammate after the next approached and congratulated him on scoring his first touchdown as a Buccaneer. Arians also embraced his new back, telling him that great things would continue to happen for him as long as he put the team first in everything he did, from the way he practiced to the demeanor he displayed on the sideline when he wasn't playing.

It wouldn't be the final conversation the coach and his running back would have on that very subject.

After nearly two decades of piloting the New England Patriots' offense, Tom Brady had to learn a new offensive language, a new playbook, and a new philosophy in Tampa. It was at times a struggle.

In 1983 Bruce Arians, age thirty-one, was one the youngest head college football coaches in America. He guided the Temple Owls for six seasons. Nearly four decades later, his staff at Tampa calls itself "Temple South," because six assistants have a connection to Arians when he coached at Temple.

Married in 1971, Bruce and Chris Arians honeymooned at a roadside hotel in Staunton, Virginia. Since then, Chris has been Bruce's closest confidante. Here the grandparents are pictured with two of their grandchildren.

Two former bartenders, two close friends: the relationship between Arians (right) and Bucs' general manager, Jason Licht, is unlike any other in the NFL.

The former head coach of the New York Jets, Todd Bowles played safety for Arians at Temple. He is the defensive version of Arians: aggressive, risk-taking, and fearless.

Bruce Arians took the Tampa job on one condition: that he relinquish play-calling duties. He turned that responsibility over to offensive coordinator Byron Leftwich, the former quarterback who played in the Arians offense at Pittsburgh and interned for him when Arians coached the Cardinals. Leftwich formed a close relationship with Brady, who is two years older than his OC.

In his first conversation with Arians during free agency, Brady told his soon-to-be-coach that his good friend Rob Gronkowski was ready to come out of retirement. Based on Brady's recommendation, the Bucs traded for Gronk, who ended up scoring two touchdowns in the Super Bowl. "That was total blind faith in Tom," said Arians.

After being released by the Jaguars, running back Leonard Fournette received a text message from Brady, encouraging him to join the Bucs. Days later, Fournette was in a Tampa Bay uniform—and would be a key player in the team's Super Bowl run.

By giving up play-calling duties, Arians was able to spend more time with his defense and special teams. "For the first time in my career I am a true head coach," Arians said. "Now I can focus on all aspects of the team rather than just the offense. I've hired coaches who don't need coaching. So I can really look at the big picture."

No one person spent more time in Brady's ear than quarterbacks coach Clyde Christensen. Before the Bucs pursued Brady in free agency, Christensen analyzed every throw that Brady had made in the previous four years. His report to Arians: "Tom has still got it."

Two months after Arians nearly cut Leonard Fournette—"He had disconnected from the team," Arians said—the running back was one of the most valuable Bucs in the playoffs. In the Super Bowl against the Chiefs, "Playoff Lenny" rushed for eighty-nine yards on sixteen carries and scored one touchdown.

For weeks Brady lobbied Arians and Licht to sign free agent wide receiver Antonio Brown. The pair finally relented and Brown—after staying at Brady's mansion for a brief stint—developed a quick rapport with his quarterback. In the Super Bowl, Brown caught a one-yard scoring pass from Brady, even though Brown ran the wrong route.

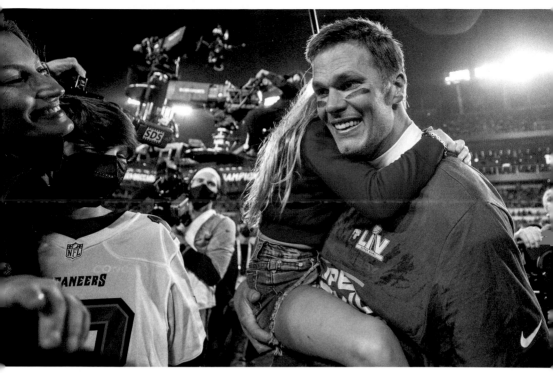

Brady savored his record-setting seventh Super Bowl victory with his children and wife, Gisele, who in the aftermath asked her husband, "What else do you have to prove?"

After forty-six years in coaching—including stops at Virginia Tech, Alabama, Temple, Mississippi State, Kansas City, New Orleans, Indianapolis, Pittsburgh, and Arizona—Bruce Arians scaled the summit of the sport, becoming the oldest head coach (68 years and 127 days) in NFL history to win a Super Bowl. As always, Chris was by his side.

Fueled with liquid courage, Brady attempted his most dangerous pass of the season, tossing the Vince Lombardi Trophy over the eighty-foot-deep waters of the Hillsborough River to a boat filled with wide receivers and tight ends. Cameron Brate made the grab. "That was the best catch of my life," Brate said. "But if I had dropped that? I think I would have had to retire."

CHAPTER 8

||||||||||||||||||||||||||||||||||||||||||||||||||||||||||||||||||||||||||||||||||||||||||||||||||

The Mastermind

He strolled along the sideline on Empower Field at Mile High in Denver, Colorado, his eyes bright with curiosity. These brown eyes have been the most useful tools in the career of Jason Licht. When he was a walk-on offensive guard at the University of Nebraska from 1989 to 1991, he closely watched NFL scouts who came to practices to identify and assess players with pro potential, noting how the scouts examined everything from ankle movement to hip flexibility to knee dexterity.

He saw the scouts monitor how each player communicated with his teammates on the field, how each reacted to getting beaten by his opponent, and how each conducted himself when he played well. Licht knew he didn't have the talent to play in the NFL, but the idea of becoming a scout intrigued him. *Maybe one day*, he thought during one Nebraska practice when a half-dozen scouts were in attendance, *I'll give scouting a try.*

Now on the sideline in Denver on September 27, Licht's eyes darted from one Buc player to the next as they warmed up before kickoff. For a few minutes he trained his attention on Brady, and he still marveled at how far the quarterback had come from when he first met him a few days after the 2000 draft. At the time, Licht was working in the scouting trenches for New England, covering the Southeast as a college area scout. Though he wasn't even sure if the head coach knew his name, Licht was in the draft war room when Belichick made the decision to draft Brady in the sixth round, even though New England had just signed quarterback Drew Bledsoe to a record ten-year, $103 million contract. "I've expressed over and over again my desire to play my entire career with the New England Patriots," Bledsoe said after the contract was announced. "It looks like that is a very real possibility."

Licht talked a few times to Brady during his rookie training camp, chatting in the hallway or cafeteria about how his transition to the NFL was going. Charlie Weis, the Patriots' offensive coordinator, was hard on Brady during his first camp, constantly yelling at his rookie quarterback for making one mistake after the next. "Then we'd watch the practice on film as a staff," Licht said, "and there were times when Charlie was like, 'Oh shit, Tom was right. I shouldn't have yelled at him there.'"

New England kept four quarterbacks on its roster in 2000, a rarity in the NFL. Brady was the last player to make the cut. "We didn't want to put Tom on the practice squad because anyone could have signed him," Licht said. "So as difficult as it was to do, we went with four quarterbacks."

Licht left the Patriots organization in 2003 to become the Eagles' assistant player personnel director. For years, his enduring image of Brady was Weis shouting at him while Brady struggled to learn a new offense. Six years later, Licht returned to the Patriots in 2009 as the director of player personnel. He first spotted Brady in the team cafeteria. Brady—who now possessed three Super Bowl rings—was talking to a few teammates, who sat in rapt attention in front of their quarterback. Licht leaned in closer and heard Brady discussing an apartment building, one that he had just bought in Boston. The former unknown young quarterback had grown up—and so had his bank account.

||||||||||||

Licht continued to walk around the grass field at Mile High in Denver before the third game of the 2020 season against the Broncos, those brown eyes inspecting the team he had put together. The Bucs were 1–1 and searching for their first road win, and Licht expected a breakout game from linebacker Shaq Barrett, who Licht had signed in the 2019 offseason to a bargain-basement one-year, four-million-dollar contract. Barrett, a former undrafted free agent from Colorado State who started his career on the Broncos practice squad, spent his first five seasons in Denver as a rotational edge defender, totaling eight sacks. He played behind All-Pro linebacker Von Miller and then in 2018 the Broncos drafted Bradley Chubb out of Georgia with the fifth overall pick to play opposite Miller, which meant Barrett wouldn't be starting anytime soon. When Barrett became an unrestricted free agent, Licht didn't focus on his low sack production; he watched hours of tape and determined he could be an ideal edge rusher in Todd Bowles's blitz-heavy defensive scheme.

On film, Licht saw that Barrett had learned several different pass-rushing techniques and moves from Miller, including the "dip-and-rip": a maneuver in which he blasted around an offensive tackle from his two-point stance, got low to the ground (the "dip"), used his inside hand to "rip" the tackle's hands off him, and then charged at the quarterback. Licht also noted that Barrett was an all-effort, all-the-time player; his internal motor never stopped. Licht concluded that Barrett was a young, dedicated, ascending player who had the potential to be a high-level starter. After the 2018 season, Denver never even offered him a deal to stay. And what a signing he was for Licht and the Bucs: In 2019 Barrett led the league in sacks with 19.5. The potential that Licht had seen on tape had transformed into production under Bowles, who in 2019 overhauled Tampa's zone 4–3 defense into a blitzing, 3–4 one-gap scheme that perfectly fit Barrett's skill set.

From his spot on the sideline, on a mild afternoon in the foot-hills of the Rocky Mountains, Licht's eyes followed the ball on the opening kickoff as it sailed through the baby-blue sky. He was especially happy because his mother in Lincoln, Nebraska, was watching the game, a treat for her; most Bucs games the pre-vious seasons hadn't been broadcast locally in the Midwest. She told her son she'd be looking for him in the stadium.

||||||||||||

Arians and Licht first met in January 2013, and they immediately hit it off. Arians, who had spent the 2012 season as the offensive coordinator and interim head coach of the Indianapolis Colts, had flown to Phoenix to interview for the job of head coach of the Arizona Cardinals. On the eve of Arians's formal sit-down with owner Michael Bidwell and key Cardinal front-office personnel, Licht drove to the Marriott Phoenix-Resort Tempe at The Buttes to meet with this new coaching candidate. Licht, then the vice president of player personnel for the Cardinals, sat down with Arians at the hotel bar.

They discovered they had much in common. They both played college football for father-like coaches—Arians a quarterback at Virginia Tech for Jimmy Sharpe and Licht an offensive lineman at Nebraska for Tom Osborne. They both bartended during their school years to make extra money. And they both had a frightening late-night experience while working in a smoky saloon that forced them to reevaluate their career paths—and their aspirations in life.

After graduating from Virginia Tech, Arians was slinging drinks behind the bar at a basement nightclub in Blacksburg, Virginia, when a man who lived nearby in a cabin in the Blue Ridge Mountains sauntered through the door, a storm brewing in his eyes. With his Rip van Winkle–like beard, the man looked like he was straight out of the cast of the movie *Deliverance*.

"Tonight," the man declared to Arians, "I'm going to drink and I'm going to fight."

"Well," Arians replied, "let's make the beer free for you, but go fight somewhere else."

A few hours passed. Then the man, filled with liquid fire, started pinching the posteriors of several young women. Arians told him he had to leave. The mountain man then pulled out a black handgun and jabbed it into Arians's stomach. "Throw me out now," he said.

Arians was terrified. But then the nightclub owner, wielding a blackjack, clubbed the man over the head, knocking him out cold. "I never miss," the club owner said to Arians.

"Well, the damn gun wasn't pointed at your stomach!" Arians replied. That was Arians's last night of bartending, at least outside of his own home or his postgame tailgate parties. "I realized that coaching would be a better career path," Arians said. "My wife agreed."

Licht, oddly, had a similar story. After graduating from college and while studying for the Medical College Admission Test (MCAT), Licht bartended and worked as a bouncer at The Brass

Rail, a watering hole in downtown Lincoln, Nebraska. One night Licht denied entry to an intoxicated young man. After the bar closed at 2:00 a.m., Licht walked through the darkness a few blocks to his 1981 red Jeep that—even in the deep freeze of a Midwestern winter—he drove with the top down.

But before he reached his car, the man who Licht hadn't let into The Brass Rail appeared in front of him, brandishing a knife. "You think you're tough now," the man said, then lunged at Licht. But the man lost his grip on the knife and, at that moment, one of Licht's former Nebraska teammates came upon the scene. The two of them wrestled the man to the ground. The outcome did not end well for the attacker; the white snow was stained with bright red blood.

"My big takeaway from that experience was that I needed to find a new line of work," Licht said. "A few months later I got into scouting. I'd say football scouting is a much safer work environment."

|||||||||||||

On the night Arians and Licht met in the Arizona desert, Arians dropped his first F-bomb about three minutes into their conversation. Later that evening Arians and Licht got together with Bidwell and Steve Keim—the Cardinals general manager—for dinner at a restaurant called Tarbells, a Zagat top-ten restaurant in the Valley. "I hope Bruce doesn't cuss as much as he did when we were together," Licht told Keim. But Arians, who has never written a résumé, much less typed out the formal three-year franchise plan that most head coaching candidates would bring to such a meeting, let loose. Only minutes after sitting down, Arians asked aloud, "Where the fuck are our drinks?"

The moment was quintessential Arians. But Arians's propensity to march to his own unique drumbeat and his lack of verbal finesse probably cost him head coaching opportunities earlier in his career. When Keim and Licht later asked Bidwell what his first impression of Arians was, Bidwell laughed and quickly said, "It's impressive how he can use the word 'fuck' as a noun, an adjective, and verb in a single sentence."

Two days after their dinner, Bidwell hired Arians to be Arizona's head coach. After Cardinal home games, Arians held tailgate gatherings from the back of his truck, serving drinks to coaches and players and front-office personnel, including Licht. The two grew even closer at these makeshift parties where Arians reprised his long-ago bartender days. Said Arians, "Win or lose, we booze, baby."

||||||||||||

Licht is as pure a Nebraskan as a stalk of corn. Born in the town of Fremont (pop: 26,437), Licht became a Cornhusker fan soon after he took his first steps. His parents, Ron and Karen Licht, moved to Yuma, Colorado, when little Jason was five, but that didn't diminish his love for Tom Osborne and Nebraska football. He watched every Cornhusker game on the television in the living room, mesmerized by the action. "I would be sick on Mondays if Nebraska lost," Licht said.

At Yuma High, Licht grew into a 220-pound linebacker. As a senior in 1989, he received a handful of scholarship offers from small schools, but turned them down the moment a Cornhusker coach told him he could walk on at Nebraska. In Lincoln, he switched positions to guard, ate and lifted weights like he was in the race of his life to gain as much weight and muscle mass as possible, and earned a spot on the scout team. In practice, he often went one-on-one with the likes of All-Americans Christian Peter and Trev Alberts. By 1991, he'd worked his way up on the depth chart and backed up future NFL Hall of Famer Will Shields and his teammate Brenden Stai. That season, as a third-year sophomore, Licht earned his first letter.

As an offensive lineman, Licht had to learn the characteristics and traits of nearly every position group on the field, which gave him a wide-ranging perspective about the talent it took to flourish. He studied the strengths and weaknesses of his fellow linemen because he needed to know what he might have to help them with if they got beat. He examined the running backs and the quarterbacks, dissecting their tendencies so he could understand how best to block for them. He watched hours and hours of film on opposing defensive linemen and linebackers, analyzing what they did well and what they did poorly. He even monitored opposing defensive backs, seeing how often they blitzed, how they blitzed, and who was most vulnerable in coverage—important because if that beatable defensive back was lined up on his side of the field, he needed to create a passing lane for his quarterback. Licht didn't realize it at the time, but he was already preparing for a career in scouting.

But Licht also scouted himself, and he was a realist. "I didn't have the genetic makeup of the guys in front of me at Nebraska," he said. "There simply were things that they could do physically and athletically that I couldn't. I knew I'd never really get any serious playing time at Nebraska, no matter how hard I tried in the weight room or at practice."

He stayed in Lincoln but transferred to Nebraska Wesleyan, a Division III school located a few miles across town. He switched to defensive tackle and led all Plainsmen defensive linemen in total tackles in 1992 and 1993; in both seasons he was named to the All-Conference team. He spent many nights bartending and working the door at the Brass Rail, a downtown dive on "O" Street. Licht was known as one of the nicest guys at the bar— "Nebraska nice," even for Nebraska—but his gentle demeanor didn't stop him from breaking up more than a few booze-fueled fights. No one who worked at the Rail during this time remembers Licht ever losing his temper.

Licht, a premed major, wanted to become a small-town doctor. Then during his senior year, a scout from the Colts traveled to Lincoln and worked out Licht. Instead of the scout interviewing Licht after the session, it was the sweat-soaked Licht who peppered the scout with questions: *What's your job like? What makes a good scout? How often are you on the road? What are you looking for in individual workouts?*

In 1995, as he was studying for the MCAT, he still lifted weights and trained at Nebraska's Memorial Stadium with former Cornhusker teammates. One day he bumped into Charlie McBride, the Cornhuskers defensive coordinator. The two had been close when Licht played at Nebraska. "What are you doing?" McBride asked.

"I'm waiting to take the MCAT, again," Licht said, smiling. "But I was thinking of trying to get into scouting."

The next day McBride called Licht. "Get to my office now," McBride said. "I've got a guy here from the Dolphins you need to meet."

That guy was Tom Braatz, Miami's director of college scouting and the former general manager of the Packers and Falcons. Braatz was in Lincoln to scout several Nebraska seniors.

Licht and Braatz hit it off immediately. The two met for drinks the next day at the Brass Rail, where Braatz explained exactly what it took to be successful as a scout and how it could be a stepping-stone into a front-office career in the NFL. A week or so later, the Dolphins hired Licht to be a scouting intern for head coach Don Shula.

Licht performed every type of grunt work, and he loved it. If he was asked to get coffee and donuts, Licht bought the best coffee and donuts he could find. If he was tasked with making highlight tapes of different college players, he would stay as late as it took so that the tapes were the best they could be. He drove players who flew into Miami for physicals from the airport to the doctor's office, always flashing his easygoing, never-met-a-stranger Midwestern personality. He escorted other players to lunch, where his conversation skills put them at ease. During games, he'd sit in a coaches' box high up in the stadium. Still photos of different alignments that the opposing team was using would arrive in the box, and then Licht would race them down to the coaches and players on the sideline. He relished every aspect of his job.

In 1998 the Panthers hired Licht as a full-time scout. Then New England offered him a job as a college scout in 1999. Before his first draft meeting with Belichick, Licht made a decision: he was going to give honest answers about players he had scouted, even if those opinions ran counter to what more tenured scouts and team personnel in the room thought. When it was his turn to speak, Licht gave thorough assessments—and they did often conflict with prevailing opinions in the room. But Belichick loved this brave, brazen young scout, because he was thorough, he was honest, and he was completely unafraid. Belichick soon promoted him to national scout and then to assistant director of player personnel.

Licht gained a wealth of knowledge from Belichick, some simply astonishing. One time he returned to Foxborough after being on the road for a month during the season. He bumped into Belichick in the hallway. The head coach asked him about a player from North Carolina. "Man, I really like him," Belichick said. "He's a high pick." Licht gave his take on the player, and then Belichick asked. "Yeah, but did you see him play against Marshall? What the hell happened to him in that game?"

Licht was floored. When in the world did Belichick have time to watch a fourth-round prospect play against Marshall? The moment only reinforced to Licht that he needed to be as prepared as anyone in the building to talk about a player if he was ever going to get promoted again. With Belichick, the planning for the future never stopped. After the Patriots upset the Rams, 20–17, in Super Bowl XXXVI on February 3, 2002, Licht found Belichick. "Wow, that was awesome," Licht said. "Now what?"

"Now what?" Belichick said with a hint of annoyance in his voice. "We win more."

Licht went back on the road, traveling the country during college football season. "I fell in love with the scouting life," Licht said. "You go to a college town, talk to coaches, then go to the game on Saturday. I was young so I'd hit the bars after the games and just talk to people and have fun. Then I'd write up my evaluations on the players. I knew they had to be incredibly exhaustive, because Coach Belichick was going to read every single word that I wrote. But I couldn't get enough of scouting."

Belichick had profound respect for Licht. After Licht left the Patriots in 2003 to join the Eagles (where he worked until 2007) and then the Cardinals (2008), he bumped into Belichick at the 2009 NFL Combine. Belichick rarely rehires anyone who leaves his organization—it is typically viewed as an act of Benedict Arnold–level betrayal—but Belichick asked Licht, "How about if you came to the combine as a Cardinal and left as a Patriot?"

New England hired Licht as their director of player personnel. His second stint with the Pats lasted four years, until he moved on to Arizona in 2013.

By 2014, Licht's résumé was impressive: he had been a part of putting together teams that had played in the Super Bowl with the Patriots, Eagles, and Cardinals. A people person, he had developed relationships with Arians, Brady, Gronkowski, and hundreds and hundreds of other coaches and players across the NFL. The Bucs hired Licht—who had the endorsement of virtually everyone he ever worked with—in January 2014 to be the team's general manager.

Licht's first five seasons were up and down—his successes in roster building often overshadowed by moments of acute failure. Licht fired coach Lovie Smith (who was hired right before Licht) after just two seasons. He then elevated offensive coordinator Dirk Koetter to head coach, who he fired after three seasons, including back-to-back 5–11 campaigns. The franchise quarterback Licht had selected with the number one overall pick in 2015, Jameis Winston, never developed into the player Licht thought he'd become.

But the most egregious draft error occurred in 2016, when Tampa Bay traded up in the second round to select a kicker, Roberto Aguayo. There is some dispute about whose decision the Aguayo pick ultimately was—"It absolutely was not Jason's call to pick a kicker in the second round," said one Tampa Bay insider with knowledge of the situation—but there is no question the selection was a disaster. Aguayo didn't last a full season. After five seasons, the Bucs' record under Licht stood at 27–53, prompting talk across the league that Licht was just another former Patriot who couldn't win without Belichick by his side.

But slowly, Licht had been building a strong roster. He had

plucked some gems from drafts: Mike Evans with the seventh overall pick in 2014; two cornerstone offensive linemen in the second round of 2015 in Donovan Smith and Ali Marpet; another talented wide receiver, Chris Godwin, in the third round of 2017. Plus, he discovered tight end Cameron Brate, an undrafted free agent out of Harvard, in 2014. He orchestrated a trade with the Giants for defensive end Jason Pierre-Paul for a third-round pick and a fourth-round pick swap, and later he re-signed Evans, Brate, and linebacker Lavonte David while still staying under the salary cap. In 2018 he added core players Ryan Jensen (a center), Vita Vea (defensive tackle), Carlton Davis (cornerback), Ronald Jones II (running back), Alex Cappa (guard), and Jordan Whitehead (safety).

Then in January 2019 Licht cut his most important deal: he signed Arians. In Arizona, Licht had worked directly with Arians on player acquisition and contract negotiations, and the two would do the same in Tampa. By this time, Licht had learned from his mistakes. He listened to his staff more and more, fostering a "team" approach to scouting and roster decisions. When Arians arrived, Licht wanted his new head coach to become a key figure in draft evaluations. At their first draft together, in 2019, Licht often deferred to Arians. They had some explosive, expletive-filled arguments, but Licht and Arians both say they "even argue well" to get the result that they want.

With Arians and Licht jointly overseeing their first draft together in 2019, the pair landed an impressive haul. Linebacker Devin White, drafted fifth overall, would soon become one of the most athletic, feared middle linebackers in football. Cornerback Sean Murphy-Bunting, their second-round pick, would play the best football of his life during the 2020 postseason. Safety Mike Edwards, taken in the third round, would also play a vital role in Tampa's Super Bowl run. And wide receiver Scotty Miller, their sixth-round pick, would become a trusted long-range target of Tampa Bay's *other* noted former sixth-rounder—and make perhaps the most important catch of the 2020 season. That offseason Licht also signed free agent defensive tackle Ndamukong Suh, who was still a force in the middle at age thirty-two.

But what was the most definitive measure of the quality of the roster that Licht had personally constructed? That came in March 2020 when Tom Brady picked the Buccaneers over the Patriots. In the end, Licht's handiwork had been responsible for signing the most successful quarterback in NFL history, and outdoing his former mentor Bill Belichick in the process.

||||||||||||||

Licht's biggest fan was his father, Ron, who owned a construction company in Lincoln, where his parents moved during Licht's college career. Ron was such an ardent Nebraska fan that his friends called him Husker Ron. He frequently traveled to Tampa Bay, where he would hang out outside of the locker room after games and take pictures with the players. From his father, Licht inherited his quick smile and gentle, easygoing personality—the type of demeanor that would have made him an excellent doctor with a pitch-perfect bedside manner.

Licht talked to Husker Ron almost every day. Ron was the oldest of five children and worked on the family farm as a kid, which meant he couldn't play sports. But he lived vicariously through his son. Then Licht's dad—his best friend—died unexpectedly during the 2019 season on a Saturday before the Bucs played the Rams in Los Angeles. Licht immediately flew home, heartbroken.

Throughout the 2020 season, Licht dearly missed his dad.

And now that he was standing on the sidelines in Denver as the Bucs played the Broncos—and only 140 miles from the childhood home in Yuma where Licht had thrown a football in the backyard with his old man on so many occasions as a kid—more images and thoughts of his father flowed in Licht's memory bank. The son was more determined than ever to make his father proud.

Out on the field, one of the players Licht had signed was dominating the action: Shaq Barrett. From his outside linebacker position, Barrett had a pair of sacks, including one for a safety, in the stadium where he had spent the first five years of his career. Defensive coordinator Todd Bowles confused Denver quarterback Jeff Driskel all afternoon with a multitude of blitzes that involved linebackers, safeties, and cornerbacks. The Bucs finished with six total sacks in the 28–10 win. "We built this team on defense," Arians said. "Tom was just icing on the cake. When we came into the offseason last year, we said, 'Hey, let's keep this defense together because it can be special.'"

And they did. Three weeks into the 2020 season, the Bucs were 2–1. For the first time in a nearly a decade, Tampa Bay had won back-to-back games by two touchdowns or more. Brady had his best game to date as a Buc: he finished 25 of 38 for 297 yards with three touchdowns and no interceptions. Still, Brady knew there was so much work to do for the offense to perform at the level he wanted.

"We're getting there," he said. "It's a long process. This would have been our third preseason game. There's a lot to learn, a lot of room to grow."

Arians and Licht also both believed that the offense was still weeks—or maybe even longer—from playing its best football. Still, this game was a step forward. Rob Gronkowski, who had been used mostly as a blocker in the first two games, caught six passes for forty-eight yards. Mike Evans snagged two touchdown passes. Improvements were being made. Timing was getting better. But, as the team flew back to Tampa Bay a few hours after the game, questions lingered.

How long would it take for the Tampa offense to run at full capacity? Would it be good enough this season for the Bucs to contend for a spot in the playoffs and possibly make a run deep into January and February?

Neither Arians nor Licht knew for certain. But during the night flight eastward, sipping on adult beverages, they liked the direction the team was heading.

CHAPTER 9

Learning and Growing

The quarterback coach drove toward the massive English-styled mansion on Davis Island, still a short distance away. Clyde Christensen admired the sparkling blue water of the bay, where the rippling surface shimmered and sparkled like diamonds in the Florida sunlight. Minutes later he spotted the Brady residence, located at Bahama Circle and Baffin Avenue. Christensen parked in the drive-through portico that keeps guests dry rain or shine and rang a doorbell. Brady's wife, Gisele, greeted him. "Coach, do you want some chocolate cake?" she asked. "You need to have your sweets."

"That's very kind, but no thank you," Christensen said. "I'm just here to talk a little business with your husband."

Brady and Christensen eventually headed to the garage, where Brady had set up his personal football laboratory. Along with about a hundred footballs, the garage contained several different types and brands of leather treatment products (including a variety of leather cleaners and conditioners), wire brushes, and a drying rack that Brady had purchased on Amazon. Each week, Brady spent hours in his garage preparing up to twenty-four game balls, a football scientist conducting his research.

The NFL allows teams to remove the waxy, slick gloss from game balls. Most starting quarterbacks prepare their game balls, but Brady—who never does anything half-heartedly—took it to another level. He was as meticulous as any quarterback in the league at getting his footballs to feel and perform as he wants, practically treating them like his little children who need to be primped just so for a Sunday birthday party.

Brady has always wanted to know how footballs react to different weather conditions. Over the years he's put footballs in clothes dryers, laid them in the sun for hours, covered them in snow, and let them sit outside in a rainstorm. Now, like a white-coated scientist, he wanted to see how the balls reacted to the humidity that is common in Tampa Bay in the summer and fall. "He used those balls left out in the humidity to really prepare," Christensen said. "The weather in Tampa is a little different from the weather in New England."

Before the season started, Brady had a serious talk with the team equipment manager about how to prepare the game balls. Brady was careful with his language; the last thing he or the Bucs wanted was any controversy with the game balls. Brady had been down this path with Deflategate—the allegation that he ordered the deflation of footballs used in the Patriots 2014 playoff victory over the Colts, which ultimately caused Brady to be suspended for four games. So now with the Tampa equipment manager he was specific in his instructions. "Here's the brushes we use," Brady said, explaining the routine he used for nearly two decades with the Patriots. "Here's the brush we use if it's going to be warm.

Here's the brush we use if it's going to be cold. Here are the fluids that work good in humidity. Here's one I found on Amazon that for some reason makes the ball real nice and tacky."

Christensen had spent years with another quarterback who was obsessive about footballs: Peyton Manning. On Saturdays after walk-through practices, Manning would wash his hands and then walk into the equipment room with the equipment manager. He'd shut the door so he could select his twelve game balls in silence, like it was some sort of ritual. He'd rub his hands over each ball. When he liked the feel of one, he'd toss it to the equipment manager and say "game." When he didn't like the feel of one, he'd say "pregame" and rifle it to the other side of the room.

Apart from the balls, Brady had one other unusual obsession: wind. He always wanted to know the forecasted wind conditions for the upcoming game. Cold temperatures didn't bother Brady—a career spent in New England had hardened this California native—but the wind was the one weather element that Brady couldn't get enough information about. He wanted to know precisely what direction the wind would be coming from, the exact speed of the wind, whether the wind speed would fluctuate during the game, at what rates and times it would change, and whether the wind tended to swirl in the stadium they were slated to play in next. No detail about the wind was too minute for Brady.

"Tom is a perfectionist," Christensen said. "And if you ever want to know anything about the wind in Tampa Bay on a game day, he's your guy. He can tell you more about it than our local meteorologist."

As Christensen left Brady's house, his quarterback was still in the garage, laboring away on his footballs, preparing for the next practice, the next challenge, the next game.

||||||||||||

On a humid autumn afternoon, with only a whisper of wind fluffing the mural-sized images of Brady and Mike Evans hanging on the exterior of Raymond James Stadium, Brady wrote another did-he-just-do-that kind of chapter in his book of performances on October 4 against the Los Angeles Chargers. The Bucs began the game with a 10-play, 75-yard drive—featuring two throws to Evans for a combined 41 yards—that ended with Brady hitting tight end Cameron Brate for a 3-yard touchdown pass. The opening play script, as written by Brady and Leftwich, had once again worked to perfection.

But the Chargers stormed to a 17-point lead in the second quarter behind the stellar play of rookie quarterback Justin Herbert, who connected with wide receiver Tyron Johnson for a 53-yard touchdown and with tight end Donald Parham Jr. for a 19-yard scoring strike. Brady, for his part, threw another pick-six, his second in four games. The errant throw underscored that the eleven players on the Bucs offense were still a touch off-key, still not playing in perfect harmony.

With 1:37 left in the second quarter, the Chargers held a 24–7 lead. Then Suh stripped the ball away from running back Joshua Kelly and Devin White recovered it at LA's six-yard line. Three plays later, Brady found Evans in the end zone, cutting the deficit to ten on Brady's second touchdown pass of the first half.

Brady was heating up. On the Bucs' first possession of the second half, he guided the offense 69 yards down the field to a touchdown, finishing with a 28-yard strike to tight end O.J. Howard. The next time the Bucs had the ball, Brady quickly mauled the Chargers defense once again: first he hit wide receiver Scotty Miller on a high-arching 44-yard-deep ball and then, on the next play, he connected with Miller again for a 19-yard touchdown throw. It was a two-play drive—the kind of aggressive offensive play calling that is the hallmark of Arians's no-risk-it, no biscuit philosophy.

But on the final play of the third quarter, Herbert threw a 78-yard strike to Jalen Guyton to give the Chargers the lead, 31–28. On the sideline, Brady walked from one offensive player to the next, fire in his eyes, yelling, "We got this! We got this!" He then led the Bucs straight down the field on a 75-yard drive that culminated in a 9-yard touchdown pass to rookie Ke'Shawn Vaughn. The Bucs added a field goal to win 38–31, pushing their record to 3–1.

How good was Brady, who finished 30 of 46 passing for 369 yards and tied a team record with five touchdown throws? For an NFL record thirty-fourth time, he had overcome a deficit of ten or more points and shepherded his team to victory. But it was more than that. He was singularly responsible for making his teammates believe they would win, and this psychological jolt of confidence that the mere presence of Brady injected into everyone on the Tampa sideline was just as important as anything he did on the field. "If this had been last year, we would have lost by thirty points," Arians said. "I guarantee you that. But we have a quarterback who gives us an assurance that no deficit is too big to overcome."

"How can you not believe in him?" Bucs wide receiver Scotty Miller said. "He's the greatest to ever do it."

||||||||||||

But Brady wasn't perfect. The next week against the Bears at Soldier Field, on a Thursday-night game in prime time, the Bucs faced a fourth and six at their own 41-yard line with 38 seconds remaining. Tampa trailed 20–19, and now Brady needed to complete a short- to mid-range throw to keep the drive alive and put the Bucs in range for a game-winning field goal. Taking the snap, Brady dropped back to pass and then rifled the ball down the seam to Brate, but it was knocked away by Chicago's DeAndre Houston-Carson. Brady immediately held up four fingers, first flashing his hand at the sideline and then at the officials, signaling his belief that Tampa's offense had one play remaining. But they didn't. Chicago won, 20–19.

Brady said his mind was focused on one thing on that final play: yardage. "Yeah, you're up against the clock and you know you're up against the clock," he said. "I knew we had to gain a chunk, so I should have been thinking more first down instead of chunk on that play." (Months later, Brady would finally admit he didn't realize it was fourth down.)

Social media lit up Brady, as the image of him holding up four fingers circulated on Twitter, Facebook, and Instagram. Virtually everything Brady does is analyzed and dissected and rehashed and broken down by pundits from New York to Los Angeles, and when he makes a mistake, it becomes the lead story on every national sports outlet and fodder for Twitter trolls. Captions that were posted with the screen shot of Brady with his raised four fingers included: "Me when I'm naked on the street and the police ask me how many beers I've had," and "If you can't remember it's fourth down it's probably time to retire." Warren Sapp, the Hall of Fame defensive lineman who played nine years for the Bucs, chimed in and wrote: "HANG IT IN THE LOUVRE."

No, this season—this transition to Tampa—wasn't easy for Brady. He wasn't perfect. He wasn't in his physical prime. Many

NFL fans delighted in seeing him look addled late against the Bears. Brady has landed on multiple "Most Hated Players in NFL History" lists throughout his career, and as he walked off the field he was stone-faced and silent, never betraying a hint of disappointment as he clenched his jaw.

But Brady knew. He was in the throes of a challenge unlike any other in his career.

|||||||||||||

The next game against the Packers didn't start well. At home against Green Bay on October 18, the Bucs fell behind, 10–0, in the first quarter. The Packers were 4-0 and looked like the best overall team in the NFL. Then things fell apart—for Green Bay.

Packers quarterback Aaron Rodgers, facing a third and ten, threw an interception to Jamel Dean, who returned it 32 yards for a touchdown. Three plays later, Rodgers tossed another pick; safety Mike Edwards raced 37 yards before he was tackled at Green Bay's two-yard line, setting up an easy touchdown plunge by running back Ronald Jones. To that point in his career, Rodgers had thrown a total of only two interceptions that were returned for touchdowns in 193 games. Now he had almost thrown two pick-sixes in the span of four pass attempts.

The rout was on. Realizing his defense was shutting down Rodgers, Brady played cautiously, opting to throw to his checkdown receivers more than he had all season—a preview of a coming philosophical shift in the Bucs' offensive attack. He finished 17 for 27 for 166 yards and two touchdowns, including one to Rob Gronkowski, who finished with five catches for 78 yards. After crossing into the end zone for the first time that season,

Gronk performed his signature Gronk Spike—a forceful spiking of the ball into the ground that features a windmill-like wind up. According to Urban Dictionary, a Gronk Spike is "The action of forcing an object (usually a football) into the ground with tremendous force as a way of celebration or because you're f---ing hammered and felt like doing it anyway." This was at least the seventy-third Gronk Spike of Gronk's career, but he's executed countless nonfootball Gronk Spikes: a bridal bouquet at a wedding reception, a puck at a Boston Bruins game, and a can of beer at a Patriots Super Bowl parade. An MIT graduate student once studied the Gronk Spike and estimated that the ball leaves Gronkowski's hand at 60 mph, faster than a PGA Tour golfer swings his driver.

After the Gronk Spike against Green Bay, Brady was one of the first players to hug his longtime tight end and close friend. It was one of the enduring images of this near-flawless game for the Bucs, who had no penalties, no turnovers, and no sacks allowed. The Bucs' defense forced the Packers to punt on their final seven possessions. The other unforgettable snapshot of the night was the look of disbelief on the face of Aaron Rodgers, who typically is as expressionless as a mannequin. Late in the game, Rodgers's eyes were as big as twin full moons as he watched his team lose, 38–10.

||||||||||||

For weeks Brady begged. He pleaded with Licht, telling him what a good person wide receiver Antonio Brown was. He tried to sway Arians, explaining that he would take full responsibility if Brown committed a single mistake off the field. "It will all be on me,"

Brady told Arians. "I'll take all the heat." Brown referred to Brady as a "big brother," and Brady did indeed consider Brown almost like family. The two had known each other for years, but now Brown had been out of football for a year. He also faced a felony burglary with a battery charge in Florida for striking a delivery driver—and a civil lawsuit for sexual assault.

At first, Arians resisted Brady's overtures. "I'm not tearing apart my locker room," he said. "Let's see how things go."

Arians had coached Brown in Pittsburgh, where Brown had been a seven-time Pro Bowler and twice led the NFL in receptions (2014 and 2015). But by the end of the 2018 season Brown—who that year had caught 104 passes for 1,297 yards and a career-high 15 touchdowns—had become a poisonous presence in the Steeler locker room. He got into a heated argument with quarterback Ben Roethlisberger and he skipped practices before Pittsburgh's Week 17 game against the Bengals, prompting Steelers coach Mike Tomlin to bench Brown.

Three months later Brown was traded to the Raiders for a third- and fifth-round pick. But his time in Oakland was a disaster from the start. He missed ten out of eleven training camp practices because of frostbite on his feet that he had suffered during a cryotherapy session because he wasn't wearing proper protective footwear. General Manager Mike Mayock then fined Brown $54,000 for unexcused absences and missed practices. Upset, Brown engaged in a verbal altercation with Mayock and had to be held back by teammates. Finally, on September 7, 2019, Brown asked to be released from the team. The Raiders happily obliged.

That same day the Patriots signed Brown to a one-year contract, and Brady took an immediate liking to him. In his only game with New England, against Miami in Week 2, Brown caught four passes for fifty-six yards and a touchdown. But then on September 20, after allegations of sexual misconduct and alleged intimidating text messages Brown had sent to an accuser came to light, the Patriots cut him.

Now, in late October 2020, Brown talked with Licht, who immediately began grilling him about what had happened during those troubling off-the-field interactions that had led to him being cut by the Patriots. Brown detailed his version of the events. "If we sign you, Antonio, you make one mistake and you're gone," Licht said. "Do you understand?"

"Yes sir," Brown replied.

Brown then spoke with Arians. "There will be no leash with you," Arians said. "One tiny misstep and it's goodbye."

"Yes sir," Brown said.

"Your role will be what I tell you it will be," Arians said. "If our guys are healthy, you're not going to be a starter."

"Coach," Brown said, "if someone else catches a touchdown pass, I'll be the first one to congratulate him."

Two days later, Licht signed Brown to a one-year contract that included $1 million in base salary and an additional $1.5 million in performance incentives—a heavy discount from the $50 million, three-year deal with the Raiders that Brown had walked away from the previous season. When Brown arrived at the Tampa airport, Brady picked him up and drove him to the Jeter mansion, where Brown stayed for a few weeks. Brady and Brown have a deep friendship—"It goes back years," Brady said—and Brady wanted Brown to feel welcomed in Tampa. Brady knew that Brown didn't have many acquaintances in town, and he wanted Brown to stay at a place where he was comfortable and could focus on one thing: football. The quarterback and receiver would spend many hours together talking about what patterns

Brown likes to run, which ones Brady likes to throw, where Brown wants the ball thrown on different routes, and what routes need to change on-the-fly if the defensive coverage changes postsnap. They were, in short, spending the time to really get to know each other's likes and dislikes on the football field—hours that would pay dividends months down the road in the most important game of the year.

||||||||||||

A week after thumping the Packers, the Bucs flew across the country to play Antonio Brown's former team, the Las Vegas Raiders. Before kickoff, Arians wandered around the field and spoke with Raiders coach Jon Gruden, who on January 26, 2003, led Tampa Bay to its first, and only, Super Bowl title with a 48–21 victory against Oakland. Gruden had been the brain behind that championship team, and he admired the roster that Arians and Licht had put together. "You guys have got a lot of weapons," Gruden said to Arians.

It showed within seconds of the opening kickoff: Brady sprayed quick passes to every portion of the field, feathering the ball into tight windows as his wide receivers and tight ends made contested catches. In the Bucs' 45–20 victory at Allegiant Stadium, Brady threw four touchdowns to four different players (Gronkowski, Miller, Godwin, and rookie receiver Tyler Johnson) and completed 33 of 45 for 369 yards. He wasn't sacked and the Raiders hit him only once as Brady compiled a passer rating of 127.0—his best so far in the season. During the previous five games, Brady had thrown 15 touchdowns and only one interception.

On the plane home, Arians huddled with Licht near the front

of the cabin, rehashing the action. The Bucs weren't perfect through the season's first seven games—their record now stood at 5–2—but the pair loved how the team was improving each week, each practice, each play. The collective focus of the team had never blurred; Arians could feel a sense of urgency at every practice, as if every player knew they were on the cusp of reaching a rare goal, but it would require more sweat equity. And no one, it was clear to Arians and Licht, wanted to commit a mental or physical error and let down Brady.

"I'm glad he's on our team," Licht said to Arians. "It's still hard to believe sometimes that we got the GOAT out there playing for us."

|||||||||||||

Bruce Arians has one major request of his placekicker: make the kicks you're supposed to make. This means if the field goal attempt is less than fifty yards, Arians demands that his kicker split the uprights more than 85 percent of the time. If the attempt is fifty yards or longer, Arians considers it a bonus if his kicker puts points on the scoreboard. "Few things are more frustrating than having a kicker miss the easy ones," Arians said. "You gotta hit the gimmes in the NFL. Points are too damn hard to score."

Since 2015, the Bucs' kickers had not hit enough of the gimmes. Their list of flameouts over that span includes Kyle Brindza, Connor Barth, the aforementioned Roberto Aguayo, Nick Folk, Patrick Murray, Chandler Catanzaro, Cairo Santos, and Matt Gay. During the previous five years, those eight kickers had combined to make 73.7 percent of their field goal attempts, well below the NFL average of 83.9 during that stretch. So during the 2020 training camp, Licht and Arians had signed kicker Ryan Succop, the last player picked in the 2009 draft, Mr. Irrelevant.

|||||||||||||

It is one of the greatest tales in NFL lore, one that went untold for years, but one that Ryan Succop now knows well.

Rewind the clock to the spring of 1976. One afternoon Paul Salata, who played wide receiver at USC in the late 1940s and was a former tenth-round draft pick by the Pittsburgh Steelers in 1951, heard a story about a man who randomly pointed to a name in a telephone book and invited the individual to Laguna Beach for a few days of fun in the sun. Inspired, Salata, a successful businessman in Newport Beach, California, came up with his own idea: he wanted to bring the last player selected in the NFL draft to Newport Beach for a week and celebrate him as if he was the first player drafted.

"My dad is an underdog and he wanted to celebrate the underdog," said Melanie Finch, Paul's daughter and the CEO of Irrelevant Week. "His motto is: 'Do something for somebody for no reason.' And so he started Mr. Irrelevant Week."

The final player picked in the 1976 draft was Kelvin Kirk, a wide receiver from the University of Dayton. Salata had a gift for showmanship—he had bit roles in several movies, including *Stalag 17* and *Angels in the Outfield*—and he promised Kirk that he would enjoy a week like no other in Newport Beach. Salata offered to cover his expenses, throw a parade in his honor, and award him "The Lowsman Trophy," an anti-Heisman award that featured a player fumbling the ball.

Kirk, picked in the seventeenth round, quickly accepted. But there was a slight glitch in the plan: Kirk overslept and missed his flight from Dayton to Southern California.

Salata had already alerted local media about his idea and that afternoon a press conference was slated to be held in front of the local courthouse. Salata, a lover of screwball comedies, improvised: he drove to his local Safeway and asked the butcher—a lean, athletic-looking forty-two-year old—to pretend to be a wide receiver from the University of Dayton.

Hours later the butcher—chest thrust out and smiling like a wide-eyed farm boy from the grain belt—was standing in front of cameras and notepads on the Newport Beach courthouse steps. He explained how grateful he was to be in sunny California. He detailed what his life was like on the Dayton campus (a place the butcher had never seen). And he emphasized how excited he was to take his first journey out west. The press devoured his every word.

The butcher rode in the back of a convertible along a parade route, blowing kisses to the adoring crowd, telling kids, "Don't smoke, don't drink . . . Play football . . . I love you."

Later in the day the "real" Kelvin Kirk finally arrived. When no media were present, he quietly relieved the butcher of his duties as Mr. Irrelevant. The local reporters never caught wind of the ruse.

And so began an oddball tradition that Succop enjoyed in 2009, when he was feted like a king during Mr. Irrelevant Week in Newport Beach.

||||||||||||

No Mr. Irrelevant had ever won a Super Bowl, but Succop's ambitions in August didn't stretch that far. In training camp he was just trying to earn a roster spot as he competed with Matt Gay, a fifth-round pick by the Bucs in 2019. To anyone who knew Arians, Succop had the inside track. In his eleven-year NFL career Succop had been 65 of 66 on field goal attempts of 39 yards or less. He didn't have the booming right leg that Gay possessed—only 47.1 percent of Succop's career kickoffs had been touchbacks—but no matter: Succop won the job.

"I always like to say, 'You're 0 for 0,'" Succop said. "Whatever you did last week, it doesn't matter. Whatever you did the week before, it doesn't matter. The only thing I focus on is the next kick—the one that's coming up—so I try to prepare each and every week to try to go out and give myself the best chance to kick well every time that my number is called."

It was called a lot against the Giants at Metlife Stadium on November 2. The *Monday Night Football* game wasn't a thing of beauty—it was more Pollack than Picasso—but Succop was perfect, connecting on field goals from 37, 40, 43, and 38 yards to help the Bucs hold on and defeat New York, 25–23. The Bucs were now 6–2, tying the franchise record for the best eight-game start (in 1979 and 2002). "Nothing fazes Ryan and we're going to need him to keep making those kicks," Arians said. "He's another weapon for us."

But flying back to Tampa late that night, Arians worried that his team was tired. The offense wasn't sharp against the one-win Giants, and the defense had surrendered a thirteen-play, seventy-yard touchdown-scoring drive late in the game that could have tied the score if rookie safety Antoine Winfield Jr. hadn't inadvertently collided with Giants running back Dion Lewis on the two-point conversion (a flag was immediately thrown but then picked up). Plus, the upcoming schedule was daunting: home against

New Orleans, at the Panthers, and home against the Rams and Chiefs, the defending Super Bowl champions.

Arians was trying to prevent his older players from wearing down. Brady often texted Arians and wrote, "Do you mind if I don't throw on Wednesday?" Arians reply was always the same: "Yeah, I don't care."

Arians also frequently asked Brady, "Do you want to take mental reps or do you just want to sit back?" A few times Brady replied, "I just need to sit back." But Brady wasn't used to the downtime. Early in the season he'd tell Arians, "This is the last day I ever take off." But then Brady started asking for the occasional Wednesday off.

Arians took the same approach with Gronkowski, telling his tight end, "Dude, you're not practicing on Wednesdays. I need you on Sunday. I don't need you on Wednesdays."

"Oh man, I'll be fresh!" Gronkowski said. He had never had days off before, and Arians told him if ever needed to sit out a practice to simply just ask him.

But during the next four games, the Bucs appeared anything but fresh. They lost to New Orleans, 38–3, in a Sunday-night matchup, as Brady posted a 40.4 passer rating, the third worst of his career in a regular-season game. The next week the team's flight to Charlotte on Saturday was delayed for six hours. The coaches and players were eventually let off the plane, which had a mechanical issue, and escorted into a private hangar. A makeshift meal was served while Brady chatted with Leftwich and Christensen on a bench for over an hour, reviewing the game plan, the opening play script, and how they thought the Panthers defense would react to the script. At one point during the delay a fed-up Arians, in front of the entire team, let loose on a member of the airline. "Get us a motherfucking plane that works and get us one now!" he said. "This is fucking bullshit. This is not how we travel. Get us a fucking plane!"

The airline employee hustled to remedy the problem, but seeing their coach so animated—and so angry—had a rallying effect on the team. "Everyone was feeling frustrated, and when Bruce voiced what we all were thinking, it kind of calmed everyone down," Licht said. "The players saw that Bruce was truly their leader and he was pissed off because he truly believed his players weren't being treated the way they should have been. It reaffirmed to everyone on the team that Bruce was in charge."

After arriving in Charlotte near midnight, the Bucs beat Carolina the next day, 46–23. But the following week Tampa lost to the Rams (27–24) on *Monday Night Football* and then, six days later, fell to the Chiefs (27–24). During this two-game losing streak Rob Ninkovich, a former teammate of Brady's in New England, was upset that Arians had told reporters that Brady had gotten "confused a few times with coverages" against the Rams. Appearing on ESPN's *Get Up!*, Ninkovich called for Arians to be fired. "I'm giving Tom Brady a new head coach because Bruce Arians right now, he's not cutting it," Ninkovich said. "I don't think Tom Brady gets confused by coverages . . . It's the first time Tom Brady has had a head coach throw him under the bus like this."

After seeing this segment air on ESPN, Arians spoke to Brady. "Damn Tom, get your boy off my ass," Arians said. "He's trying to get me fucking fired and he doesn't know what the hell he is talking about. He thinks we don't get along. I love it when guys talk about shit they know nothing about."

Brady laughed it off—Arians's request was made in jest—but both quarterback and coach knew something needed to change, especially on offense. The Bucs had lost three of their last four games, lowering Tampa's record to 7–5 at their bye week.

"Everybody tried to hand us the Lombardi Trophy in August," Arians said. "You don't just throw guys out there with names. You've got to practice. You've got to learn to get in sync with each other. That takes time."

Problem was, Arians and Brady were running out of it.

|||||||||||||||||||||||||||||||||||||||||||||||||||||||||||||||||||||||||||||||||||||||||||||||||||||||||||||||||

CHAPTER 10

The Turning Point

They needed to talk.

The coach and quarterback planned to play fifty-four holes at Tampa's Old Memorial Golf Club during the bye week, an open-air opportunity to dig deep into the problems that were hampering Tampa's offense. But the NFL office in New York vetoed the idea, informing the Bucs that the league had a rule forbidding players and coaches from gathering outside football facilities due to COVID-19 protocols. So Arians and Brady had their heart-to-heart over the phone.

After the Bucs lost to the Chiefs, the national media began wondering if the Arians-Brady marriage was destined to fail. Watching the Bucs' offensive struggles, former Patriots executive Mike Lombardi—one of Belichick's best friends—publicly suggested that Tampa ownership may have to cut ties with either Arians or Brady because their two offensive philosophies simply weren't meshing. The narrative that Arians (who loves to throw the deep ball) and Brady (who prefers shorter, underneath passes) weren't getting along gained momentum across the sporting landscape. Was their relationship doomed? Was it possible that Brady couldn't win without Belichick? Those questions filled hours and hours of airtime on ESPN, FOX, and virtually every other national sports outlet.

"That was all bullshit," Arians said. "Tom and I would laugh about it when we heard that people were saying we were at odds about different things. Like any coach and his quarterback, we weren't always going to see eye to eye, but we never—and I mean, fucking *never*—didn't get along with each other. Hell, he single-handedly changed the entire culture of the organization. He gave the players the confidence and belief that they could do something special because he'd done it himself so many times. Why in the hell would I ever do anything to jeopardize how he felt about being on this team? We just had a few things philosophically that we had to work on. We could have gotten these things addressed if we'd had more time together in the offseason and had a regular preseason, so that slowed us down. But the idea that we somehow were fighting with each other is just ridiculous."

Still, a series of plays during the Rams game on November 23 at Raymond James Stadium illustrated the offensive issues Brady, Arians, and Leftwich were having. With two minutes remaining in the first half, the score was knotted, 14–14. The Bucs had the ball at their own forty-eight-yard line, first and ten. In a situation like that near the end of the half, Brady wants to manage the game and the clock. He wants to drain time and score in the closing seconds of the first half, thus preventing his opponent from having enough time to score before halftime. If Brady and the offense executed in typical Brady fashion, that would essentially give the Bucs an extra possession. In New England, Brady was always taught to play situational football, meaning he needed to understand every circumstance—the score, field position, down and distance, time on the clock—before deciding what to do on the field.

But Arians and Leftwich are aggressive by nature; they want to attack, attack, attack, and they did exactly that against the Rams. On first down, Brady threw deep to Mike Evans. It was a low percentage pass that ended up incomplete, stopping the clock. Two plays later, facing a third and nine, Brady threw another pass that also missed the mark. The clocked stopped again: 1:12 remained in the half. After receiving a punt, Los Angeles moved the ball down the field and kicked a field goal—three points that ended up being the difference in the Rams' 27–24 win.

Starting with the workouts that Brady had organized in the spring, he had been practicing his deep ball throws with Scotty Miller. If the defense played man-to-man with no safety help over the top, Arians and Leftwich wanted Brady to audible out of whatever play was called from the sideline and take a shot deep with Miller on a Go route. They hit that play time and again in practice, but in games Brady's long ball accuracy hadn't been the same.

In New England, Brady preferred high-low reads in which he would always have someone crossing in front of him—a player he could get the ball to quickly on a shorter route. But this ran contrary to Arians's fondness for down-the-field routes. Figuring out how to blend these contrasting philosophies—and to fix what ailed the Bucs offense—was the subject of their phone call. It was a call that would change everything.

|||||||||||||

The November 30 call lasted more than an hour. Arians understood that to save Tampa's season things needed to change. At the time, the Bucs were clinging to the sixth seed in the NFC playoffs, but now Arians wasn't even thinking about the postseason; he just wanted to get on the same page with his quarterback. Arians believed he could use the last month of 2020 to jump-start the 2021 season, to get the ball of momentum rolling for a charge to the playoffs in the next season. During this conversation, the last thing on Arians's mind was winning the Super Bowl to be played in Raymond James Stadium in less than three months.

"We're going to figure this out," Arians told Brady. "If you don't like something in the playbook, we're going to throw it out. We need to do what you're comfortable with. If something isn't working, we'll ditch it. We're going to come together and we're going to do it right now. I need you to feel like every time we call a play, it's a play that you believe in and it's a play that will work. If you don't have that belief, then let's talk about it and we'll get it right if you think that's best."

Arians asked Brady what offensive schemes he wanted to run and how they could continue to tweak the game plans and play calls during the final four weeks of the regular season. The Arians-Brady offense had been changing since the first day of training camp—"Byron and I were constantly talking to Tom about what he wanted and how we could incorporate more of what he was comfortable with into our plan," Arians said—and this conversation marked another step in that evolutionary process.

Brady—who had preferred an offense that, in baseball vernacular, featured more small ball than a home-run-or-strike-out approach—emphasized the approach he'd followed in New England, where he used more than twelve personnel concepts that featured two tight ends or even three tight ends with thirteen personnel. As a Patriot, he also threw more passes to his running backs. Brady didn't use the phrase "more conservative" in the call, but that was essentially what he wanted the offense to become over the final month of the regular season—and, if they made the playoffs, beyond. Arians said he would always be perfectly fine with check-down throws to keep the chains moving.

"We had a melding of the minds," Arians said.

What was most important to Arians was that the offense felt natural to Brady, that he wasn't doing things on the field he wasn't comfortable with. And there was a sweet spot between the Arians Way and the Brady Way: mixing Arians's favored intermediate and deep shots with Brady's fondness for methodical, short passes to slowly move the ball down the field. Coach and quarterback both gave in a little. "I'm good with that," Brady said on the call, responding to Arians's idea that they would still take shots down the field. "We're going to start hitting more of those. We're getting there. Just need to keep working."

Brady told teammates that he had a great talk with Arians. "We're going to get things going in the right direction," he said. "I'm sure of it. I just need more reps and we need to get fresh as a team."

Brady—after his call with Arians—phoned Leftwich. "We just have to find a way to get a few wins and then we'll start playing our best ball here in December," Brady said. "We're so close. I know we are."

Brady also told Leftwich how much he enjoyed playing for Arians. Brady felt truly appreciated by his head coach—a feeling that, dating back to college, was not the most common of experiences.

||||||||||||

On many days in his adult life—and even now as he commuted to the Bucs facility in the predawn darkness and drove home in the early evening during the bye week—Brady reflected on his life and times, on how far he'd come since enrolling at Michigan. It was his reminder of the work that had to be done to achieve success; it was what drove him, all day and every day—the wellspring of his motivation.

Back in January 1995, during his recruiting trip to Ann Arbor, the young man from California was blown away by the tradition of the Michigan program and how the coaches—especially head coach Gary Moeller—appeared to believe in him. The coaching staff walked Tom through a tunnel and onto the field at Michigan Stadium, where Brady stared up, up, and up at the bleachers, which seemed to stretch into the basement of heaven. At the sight of the massive stadium, Brady was sold. "I was hoping that one day I would get to lead the team onto Michigan Stadium in front of 112,000 fans," he recalled.

A lightly recruited quarterback—his only other major offer came from the University of California, Berkeley (Cal), practically next door to his home—Brady signed his letter of intent to play at Michigan in February 1995, a few months after Moeller sat in the Bradys' living room and told the family that Tom would soon be at the center of the program, that he was the future of Wolverine football. But three months later, Moeller resigned after he was arrested for disorderly conduct stemming from an alleged drunken incident at the now defunct Excalibur restaurant in Southfield, Michigan. Suddenly, Brady's number one advocate in Ann Arbor was gone, replaced by a coach—Lloyd Carr—who knew nothing about him. When Brady arrived on campus in the summer of 1995, he was the seventh-string quarterback.

He redshirted his first year. In his second season, 1996, he was the third-string quarterback, behind starter Scott Dreisbach and Brian Griese, the son of NFL Hall of Fame quarterback Bob Griese. In the third game of that season, Carr inserted Brady late in a game against UCLA with the Wolverines leading, 35–3. Brady rifled his first pass to his left—directly into the arms of Bruins linebacker Phillip Ward, who returned the interception forty-five yards for a touchdown. It was UCLA's only touchdown of the game, and Brady was mortified. He feared he'd never enjoy the trust of his coach again.

Still buried at number three on the depth chart, Brady asked for a meeting with Carr. "Coach, I think I am going to transfer," Brady said.

"Why?" Carr asked.

"I don't think I'm getting a fair chance to play and I don't know if I'll ever play here," Brady replied.

"It'll be the biggest mistake of your life, something you will regret the rest of your life," Carr said. "You came here to be the best. You came here because of the great competition. If you walk away now, you'll always wish you had stayed and tried to compete, tried to become the best. If you leave, you'll always wonder what would have happened if you stayed."

Carr told Brady to take a day to consider his decision. Brady drove back to his off-campus apartment and called his dad, who told him he would always love him no matter what, but that he believed he should gut it out at Michigan. His father explained that kids too often run from difficult situations, that they don't display the fortitude of a champion by quitting or shrinking from competition. He told his son, in essence, to fight, to work, and to get better every day—in the weight room, in the film room, and on the practice field—and that he needed to outwork, outhustle, and outthink his competition. Decades later, Bruce Arians could have lifted these words almost verbatim when he had that long talk with his quarterback during the bye week of 2020.

The next day Brady told Carr, "I'm staying. And I'm going to prove to you that I am a great quarterback."

Brady also did something that was rare at the time: he sought the help of a sports psychologist, who emphasized to Brady that he needed to stop worrying about the other quarterbacks on the roster and focus only on what he could control. The nineteen-year-old Brady was filled with self-doubt, feeling alone and isolated from his family in California. Greg Hardin, the sports psychologist, emphasized to young Brady that he couldn't play the role of victim; it was time for him to grow up and become a man of substance and character, of integrity and honor. Brady would never forget these life-changing sessions with Hardin.

But Carr still never showed unwavering support for Brady. In 1997, Brady's redshirt sophomore year, he again was the third-string quarterback on a Michigan team that split the national championship with Nebraska, attempting only fifteen passes. Brady did finally ascend to the starting role in 1998 and set Michigan records for pass attempts (323) and completions (200) in a season, but in his senior year he split time with Drew Henson, with Carr operating a platoon system with the two quarterbacks during the first seven games of the season. Henson had missed spring practice because he was playing baseball for the New York Yankees in the minor leagues, but Brady was at the football facility every day, grinding for hours. He was voted team captain—an honor that to this day he calls the greatest of his career.

"I didn't have an easy experience," Brady told Michigan players when he returned to campus to talk to the team in 2013, after he'd won an MVP and multiple Super Bowls. "I didn't come in as a top-rated recruit. I didn't come in with every opportunity to play right away. I had to earn it."

After the sixth game of his senior season, a 34–31 loss to Michigan State in which Brady led a furious fourth-quarter comeback that fell just short, Carr named Brady his full-time starter—finally. Brady then guided the Wolverines to several fourth-quarter comeback victories, including a tour-de-force performance against Alabama in the Orange Bowl. After twice trailing the Crimson Tide by 14 points, Brady wound up throwing for 369 yards and four touchdowns in Michigan's 35–34 overtime win.

But Carr's clear ambivalence toward Brady was a red flag to NFL teams. Surely, Carr must know something, they thought. At the NFL scouting combine in Indianapolis in February 2000, Brady—with his shirt off and looking less than muscular— measured six-foot-four and weighed 211 pounds. He ran the 40 in 5.28 seconds, one of the slowest times in history for a quarterback at the combine. His vertical jump was only 24.5 inches, also one of the worst on record for a quarterback.

A post-combine scouting report about Brady was scathing. "Poor build, skinny, lacks great physical stature and strength, lacks mobility and ability to avoid the rush, lacks a really strong arm, can't drive the ball downfield, doesn't throw a really tight spiral, system-type player who can get exposed if forced to ad lib, gets knocked down easily."

"He wasn't the most impressive guy at the combine," said Arians, who was then the quarterback coach for the Indianapolis Colts. "Still, there are things you can't measure about quarterbacks and that's their heart and their grit and their determination and their willingness to work their asses off and be leaders. But hell, his own coach at Michigan didn't even seem to believe in him. Yet here's the thing about Tom: he will outwork anyone. And when your quarterback is your hardest worker, you got it made."

||||||||||||

Brady has always loved tight ends, a result of Belichick's affinity for the position. In 2000—the year Brady was drafted by New England—Belichick even had his staff jot down a mission statement. "A tight end for the New England Patriots must have functional Football Intelligence," the report stated. "He must learn the run offense like a lineman and he must learn the pass offense like a pass receiver. He must have the toughness to block a DE [defensive end] 1 on 1 (this is a common matchup). He must have good quickness. Most of his game is played against LBs [linebackers] in a short area. He must play with suddenness, getting off the line—getting in and out of cuts—getting open quickly and getting into blocks to secure the edge."

During the previous weeks, Brady and Leftwich incorporated into the play call sheet more plays that featured throws to the tight ends. Brady also self-scouted. Using his iPad, he examined hours and hours of film, identifying what plays had worked through twelve games and what concepts hadn't succeeded. He took particular care in analyzing—frame by frame, millisecond by millisecond, over and over—each of his eleven interceptions. Brady was still searching to get a better feel for the offense, to make it seem like it was his first language rather than a foreign language.

Quarterback coach Clyde Christensen studied the final year of Peyton Manning's career, analyzing his Super Bowl run with Denver and how then-Broncos head coach Gary Kubiak had melded his offense with Manning's preferences. "It wasn't easy, but Peyton got it figured out," Christensen said. "There are no magic answers. It just takes time and practice and getting to know the guys around you. But what Peyton had to do was the closest comparison to what Tom was now trying to do. There definitely was some suffering involved."

Tampa Bay fans know suffering. The Bucs hadn't had a double-digit winning season since 2010 and at the start of the 2020 season owned the worst historical winning percentage—38.5—of any franchise in any of the major four sports. After twelve games and a 7–5 record at the bye week, local columnist John Romano of the *Tampa Bay Times* seemed to speak for the masses when he called the team "the lords of the ordinary. Kings of the OK."

Other than one magical season in 2002 that ended with a Super Bowl victory against the Raiders, fans in Tampa had been waiting for a winner—a wait that stretched back more than forty years. To truly understand why the Buccaneer faithful had grown to expect the worst and, yes, had developed a collective inferiority complex, travel back through time, to 1976, the year the franchise was born and endured a season like no other in NFL history.

||||||||||||||

"Sometimes I wish it would go away," said Richard Wood, a linebacker with the 1976 Bucs. "Sometimes I wish people could forget that we didn't win a game. People don't realize how hard it was for an expansion team in 1976. But one thing I took from that season was some stories—really great stories."

One tale: Gasping for air during one of the first days of Tampa Bay's inaugural training camp in the summer of 1976, Ricou de-Shaw had a simple question for one of his coaches: Could he use the bathroom? The coach excused deShaw, a free agent tight end from the University of Miami, and he ran as fast as he had all day toward the sparkling new locker room at One Buc Place. He set his helmet down on a bench, pulled off his sherbet-colored jersey and his pads, threw all of his belongings in his locker into a duffel bag, and then—*poof!*—he disappeared, vanishing into the Gulf Coast breeze without a trace, never to be seen by the coaches or players again.

"I guess Ricou didn't want to deal with it," said Wood, who went on to play nine seasons for the Bucs and later became the head football coach at Wharton High in Tampa. "Maybe he could see what was coming."

For the Bucs who stayed, what was coming was a four-month-long seminar in the art of losing. They finished 0–14, becoming the first team since the merger of the AFL and NFL in 1970 not to win or tie a game in the regular season. "We actually weren't as bad as our record indicated," said Lee Roy Selmon, a defensive end on the team who went on to become the athletic director at the University of South Florida. "Believe it or not, we had some fun that season."

The story of Tampa Bay's woebegone debut season began on April 25, 1974, when the NFL awarded Hugh Culverhouse, a Jacksonville lawyer and real estate maven, a franchise in Tampa Bay. One of Culverhouse's first tasks was to hire a coach. Of the many résumés that arrived on his desk, none was as impressive as John McKay's. McKay had led USC to four national championships between 1962 and 1974 and, at age sixty-two, had an impressive track record for bird-dogging talent. On Halloween night in 1975, McKay signed a five-year, $2 million contract— well below what most NFL coaches were being paid. "I think we can do the job," McKay said at his introductory press conference. "It may take a little time, though."

McKay made his first trade on April 2, 1976, when he sent two players and a second-round pick to the San Francisco 49ers in exchange for the man he hoped would be the Bucs' first great quarterback, Steve Spurrier. The third pick of the '67 draft by the 49ers, Spurrier had spent most of his career backing up John Brodie, and he welcomed the trade because McKay declared him the starter before the start of training camp.

A week later, Tampa Bay used the first pick in the draft to select Selmon, a massive defensive end from Oklahoma who had won the Outland Trophy. By the time the college and expansions drafts were over, the Bucs had sixty-four players, all eager to be part of a new team in Tampa. When the Bucs held their first minicamp, a few in the organization talked of making the play-offs.

The coaches' optimism, however, started to wane as soon as they inspected some of the players they had picked in the expansion draft. "There were guys limping when they got off the bus," recalled Wayne Fontes, an assistant who later became head coach of the Detroit Lions.

The problem stemmed from how the expansion draft was conducted. Tampa Bay and Seattle—the other expansion team in 1976—weren't given the list of available players until a few hours before the draft. They had no medical reports on the draftable players and only limited scouting reports. This, plus the fact that the quick fix of free agency didn't yet exist in the NFL, soon made it clear that the Bucs and the Seahawks faced an uphill struggle. "The coaches really felt like the league didn't give them a legitimate chance to compete," said Parnell Dickinson, who was the Buccaneers' backup quarterback and later owned an insurance company in Tampa.

Even though McKay was in his early sixties, he had no NFL coaching experience—and it showed right away. The team's first preseason game took place on July 21 against the Rams at the Los Angeles Memorial Coliseum, where McKay had such success at USC. Two days before kickoff the team boarded a charter flight in Tampa, and as soon as Selmon found his seat he discovered he was sandwiched between two other burly defensive linemen on the cross-country flight. "Coach McKay decided to give seat assignments according to our positions," recalled Selmon. "Right then I wished I was a kicker."

Herewith, a quick review of a season that was decidedly not spent in the sun:

Week 1, at Houston. After finishing the preseason with a 1–5 record, the Bucs played their first regular season game on September 12 at the Astrodome. After the final whistle blew in Houston's 20–0 victory, Tampa had more yards in penalties (80) than rushing (49) or passing (59). Oilers quarterback Dan Pastorini exploited the slow-footed Bucs secondary for two touchdowns.

Week 3, versus Buffalo. A few days before the game McKay addressed the team. "Gentlemen," McKay said, "losing starts with mistakes, losing starts with turnovers, losing starts with . . ." Just then, in the back of the room, McKay spotted offensive lineman Howard Fest sleeping in his chair.

"Fest!" McKay yelled. "Where does losing start?"

Startled, Fest replied, "Right here in Tampa Bay, Coach."

The team erupted in laughter, and the incident seemed to pull the team closer together. Led by Selmon, his brother Dewey (Tampa's other starting defensive end), inside linebacker Larry

Ball, and right defensive tackle Mark Cotney, Tampa's defense stifled Buffalo's running back O.J. Simpson, holding him to 39 yards on 20 carries. With only a few minutes left to play in the fourth quarter, Tampa was in the game, trailing 14–9. The Bucs faced a fourth and ten from the Buffalo 20. McKay—the original no-risk it, no-biscuit coach—called for a fake field goal, but Spurrier's completion only went for nine yards. Tampa lost.

Week 6, versus Seattle. After road losses to Baltimore and Cincinnati, the Bucs returned home to face the winless Seahawks in a game dubbed "the NFL's game of the Weak." Said Seattle coach Jack Patera before the opening kickoff, "Hopefully one of us is going to win." Evidence of fan frustration was abundant in the stands. One sign read "Throw McKay in the Bay." Another fan wore a T-shirt that read "In Poland They Tell Buc Jokes."

The game lived up to the hype: it was indescribably pitiful. The best offensive players were the ones wearing black-and-white stripes, as the officials marched off 300 yards in penalties—more yards than either the Bucs (285 yards) or the Seahawks (253) gained all day. With Tampa Bay trailing 13–10 and less than a minute to play in the fourth quarter, the Bucs placekicker Dave Green lined up for a 35-yard field goal attempt. The snap was good, the hold was good, the kick was . . . blocked. The Bucs lost.

In his postgame press conference, McKay was asked what he thought of his team's execution. Without missing a beat, he replied, "I'm in favor of it."

Week 14, versus New England. With six seconds left the playoff-bound Patriots had the ball on Tampa's one-yard line, up 24–14. Instead of mercifully letting the clock expire, New England called a timeout. Patriots quarterback Steve Grogan was one rushing touchdown away from setting the NFL record for most rushing scores by a quarterback in a single season, and after the timeout he ran a quarterback sneak into the end zone, thus ending Tampa's 0-14 season.

"My mother told me there would be days like this," McKay said after the game. "But not every day."

As for Ricou deShaw—the Buc who former players say walked off the field never to return after using the restroom—he claimed years later that he had suffered a back injury and that was the reason why he left the facility in such haste. But the story lived on—facts be damned—because it so eloquently encapsulated the sprawling, sad tale of the '76 Bucs: a team that set the standard for NFL futility, one that would live in professional football infamy, one that would cause Bucs fans for generations to believe the worst was always yet to come.

||||||||||||

During the bye week, Arians gave the players several days off. "Go home," he said after the team had lost to the Chiefs before the bye. "Take care of your bodies, get some rest and try not to even think about football. This has been a grind. Our bye is coming twelve weeks into the season. We're all tired, and we've been playing like a tired football team. So take time off, spend time with your families, stay safe, and come back reenergized and ready to fight like hell."

"The bye week gave us a chance to breathe and relax," Leftwich said. "We self-scouted and figured out what we were doing well and what we weren't doing well. We learned a lot about ourselves during the bye. As a play caller, I felt like I had a much better idea of what Tom liked in certain situations and what he didn't. Tom and I had spent camp and the first twelve weeks of the season learning about each other and from each other, and now I sensed that I really knew what Tom was looking for in terms of play calls in whatever down and distance situation we were in."

When the players reconvened at One Buc Place to begin their preparations for the Minnesota Vikings—their first game after the bye, to be played on December 13 at Raymond James Stadium—Arians called a team meeting. Holding it outside under a vast tent, Arians knew the stakes were high: his team's moment of reckoning had arrived.

"We're some motherfuckers who are going to find a way to win a game," Arians said. "The playoffs begin now for us. We can't lose another game and expect to have a chance to win the Super Bowl. Remember what I asked you at the beginning of training camp: Do you want another team dressing in our locker room for the Super Bowl? I sure as hell don't.

"We're going to go 11–5. We're not losing another game. Let's trust each other. Let's care for each other. Let's play smart for each other. And let's bust our ass for each other. We've got too much talent right here on this team. Now let's go!"

The players rose from their chairs. They heard the prophecy of their preacher. To a man, they all knew: their season was on the brink.

|||||||||||||||||||||||||||||||||||||||||||||||||||||||||||||||||||||||||||||||||||||||||||||||||||||||||

CHAPTER 11

A Man Named Tom in Full

There was no panic, no worry, not a single outward indica- tion that something was amiss, no action or emotion that broadcast anything was bothering him. Throughout the prac- tices leading up to Tampa's game against the Vikings after the bye week, Brady appeared as steady and as in control as he had been when the Bucs had won six out of their first eight games. He still treated every offensive snap at practice like Tampa was in the fourth quarter of the Super Bowl. And now that the Bucs had lost three of four games? Nothing changed for Brady—not in the huddle, not in the film room, not in the locker room. No matter where he was, Brady had that same liquid gleam of intensity in his eyes, that same easy, confident, slow gait—a walk that metaphor- ically shouted *I got this*! He didn't engage in much small talk with other players, but that only freighted his words with even more gravity when he did open his mouth.

His teammates—not to mention the rest of the NFL—marveled

at Brady's inner drive at age forty-three. How did he do it? Most NFL players say they know it's time to retire when you start to get injured more frequently and the joy of the game, the thrill of competition, begins to dry up. Playing in the NFL is a hard job, a physical and intellectual and emotional undertaking that lasts from dawn until dusk nearly every day for six consecutive months. Yet Brady, even in his professional twilight, did not seem to have lost a smidgen of that famous will.

The Tampa players were flabbergasted by the little things he did to keep his mind and body in shape. He stretched for hours. He went to bed early, generally before 9:00 p.m. He was usually the first player to arrive at the facility. He studied each opponent like he was cramming for the most important final exam of his life. And his diet was a constant topic of wonder during the season when the players gathered for meals in the team cafeteria at One Buc Place. Some joked that he took secret weekly trips to St. Augustine, Florida—about two hundred miles away—to sip from the Fountain of Youth.

Brady has described his diet as a "mix of Eastern and Western philosophies." Once he rises from bed around 5:30 a.m., he typically will down a twenty-ounce glass of water infused with electrolytes. Then he'll have a smoothie fortified with nuts, seeds, blueberries, and bananas, often followed by a breakfast of avocado and eggs. During his early-morning workout either at home (during the offseason) or at the facility (during the season), he'll drink more water with electrolytes. Throughout the course of a day, he drinks anywhere from twelve to twenty-five glasses of water. For lunch, he prefers salads with nuts or fish, along with vegetables. He'll often have a mid-afternoon snack of hummus and more nuts and vegetables, and then for dinner he'll typically have chicken and more vegetables. The distinguishing feature of his diet is the veggies, which total more than half by volume of what he puts in his mouth. On game days he usually keeps his meals light: a morning smoothie followed by an almond butter and jelly sandwich a few hours before kickoff. He does have a guilty pleasure that he indulges in on rare occasions: avocado ice cream.

Brady has been working with his personal body coach, Alex Guerrero, for nearly two decades. Brady spends more time with him during the year than any other person, including his wife. In 2017 Brady and Guerrero traveled to Japan and Asia together for two weeks, but it wasn't your typical vacation: they held two daily two-hour practices, with Brady slinging Guerrero pass after pass; they ate prepackaged food that was organic, rich in protein, and featured no dairy or sugars; and they walked the streets in near anonymity, enjoying being American tourists abroad. Aside from a little sightseeing, almost everything they did during the trip was geared toward a single objective: figuring out how to keep Brady playing deep into his forties.

Since his early thirties, Brady has focused on injury prevention. As a younger player he was trapped within the cycle that most players endure: play the game, get hurt, rehab, and do it again and again. That was before he discovered the magic of muscle pliability and began to practice the daily low-intensity exercises that prepare his muscles to absorb and endure the physical punishments of playing quarterback better than anything else. Guerrero has designed detailed stretching routines for Brady, who receives a soft-tissue massage after each workout session. When he has time, Brady also plays games on his computer that challenge his mind—think of looking at two pictures that are nearly identical and then identifying the miniature discrepancies—to bolster his cognitive skills.

Brady has seen a great many quarterbacks lose their skills in their late thirties and forties. His boyhood idol was Joe Montana, and he watched Montana have two so-so final seasons with the Chiefs at age thirty-seven and thirty-eight before calling it quits. Brett Favre nearly led the Vikings to the Super Bowl at age forty, but after taking a brutal beating in the NFC championship game against the Saints in what became known as Bounty-gate, he faded fast and retired less than a year later. "Tom takes better care of his body than anyone I've ever met," Arians said. "He's so careful about what he eats, how he trains, and how he prepares. But trust me: he can be as funny as anyone."

Perhaps Brady's sense of humor is an overlooked reason for his longevity—*against the assault of laughter nothing can stand*, Mark Twain once wrote—because he loves a good practical joke. Once at Michigan, he and teammate Aaron Shea posed as newspaper reporters from the Brighton Bee—an outlet that only existed in Brady's imagination—and phoned freshman wide receiver David Terrell. During the "interview," Brady and Shea probed Terrell about his goals—after they had put him on hold several times—and finally got the freshman in his first week on campus to admit that his number one aspiration was to win the Heisman Trophy. Brady doubled over in laughter at the preposterousness of it: a freshman who just arrived at school *already* thinking about a Heisman. After Brady revealed to Terrell that it was his quarterback on the phone and not a reporter, he had a message for his freshman wide receiver: stay grounded, be humble, and think of *the team, the team, the team*—as former Wolverine coach Bo Schembechler once famously said. Brady didn't share the prank with the other players, because he didn't want to embarrass his young wide receiver, but a key message had been delivered.

In New England Brady engaged in several prank wars, most

notably with backup quarterback Matt Cassel. A rookie in 2005, Cassel had a habit of bursting through the door of the quarterback room. Brady told him to slow down, worried that the door would pop him in the face. But Cassel kept rushing through like he was late for a meeting, so one afternoon Brady put his foot behind the door, resulting in Cassel spilling a plate of food on himself when he tried to aggressively open the door. The battle was on.

Cassel filled up Brady's Air Force One shoes with a protein shake; Brady poured a protein shake on Cassel's head and peed on his practice jersey. Still not backing down, Cassel dumped garbage on Brady's car. On the practice field the following day, Brady said that Cassel could apologize or call him "Captain Longshanks." If Cassel didn't do one of these two things, he would suffer the consequences.

"There's no way," Cassel said.

"This is your last opportunity to apologize or call me Captain Longshanks," Brady said.

"Not going to do it," Cassel replied.

Brady didn't utter another word. After practice Cassel spotted a group of players huddled around his locker. As he walked closer, Cassel saw that there were three car tires on the floor next to his locker room chair. But not just any three tires—they were *his* tires. Cassel ran out to the parking lot and found his car sitting on cinder blocks. Brady had won the prank war. "The moral of the story," Cassel said, "is don't mess with people who have more money than you."

Arians has always liked his starting quarterback to have a sense of humor; he believes it bonds him with the other players. One of the best practical jokers Arians had ever been around was Peyton Manning. As Manning's quarterback coach in Indianapolis in

1999—Manning's second year in the NFL—Arians had a front-row seat to a practical joke that is still talked about today in the Colts organization. That year Manning's back up was Steve Walsh, who obsessively brushed his teeth multiple times a day. Wanting to play a joke on his backup, Manning one morning bought a toothbrush that looked just like Walsh's. Manning then went number two in the toilet, threw the toothbrush in, and snapped a Polaroid of it. At lunch that afternoon Manning asked Walsh if he had brushed his teeth that day. Walsh said he had. Then Manning slid the Polaroid over to Walsh, who nearly threw up. The entire lunchroom broke out in laughter.

Locker room jokes are often crude and disgusting, but they build camaraderie and unity. "Tom keeps everyone on their toes because they know his history of practical jokes," Arians said. "He enjoys messing with guys when the time is right. It's just another way for him to connect to the rest of the team."

Brady essentially had won the Buc players over the moment he signed his contract in March, but even with his elevated status as a six-time Super Bowl champion, he still longed to be just one of the guys, an easy-to-talk-to player who liked to kid around with his teammates in the locker room.

Now, though, was no time for jokes: Tampa's do-or-die moment of the 2020 season was approaching. Brady's unshakable confidence—one of many traits that made him unlike any other player in the NFL, a characteristic that differentiated him from the regular guy he so wanted to be in the locker room—never wavered. At the team's final walkthrough practice before the Bucs hosted the Vikings, Brady told Leftwich that he really liked the offensive plan they had put together. "We're getting closer," Brady said. "I can feel it coming together."

||||||||||||||

The coach didn't like what he saw.

For a few weeks, Arians had noticed that Leonard Fournette wasn't interacting with his teammates during practice. He stayed mostly to himself, standing off to the side when he wasn't lined up behind Brady with the first team offense. In the previous four games before the bye week, Fournette had carried the ball only 19 times for 47 yards, an average of 2.42 yards per rush. After his breakout performance in Week 2 against the Panthers when he gained 103 yards, Fournette hadn't rushed for more than 52 yards in a single game. He was entrenched as the backup behind Ronald Jones.

Arians, who learned to read the body language of customers during his bartending days, approached Fournette during the walk-through practice two days before the Bucs took on the Vikings. For the entire practice, Fournette had sat on a watercooler. Arians asked his running back to walk with him so the two could talk.

"I know you think you should be starting," Arians said. "But this is the situation: you're not starting. You have a role here. Right now your role is to back up RoJo. You can't sulk about it. You can't distance yourself from the team. You can't show bad body language when you're not playing. You either accept your role or I will cut you. It's as simple as that. We want you. I think we're going to need you down the road before this season is over. We're about to go on a run here and your role could change at any moment. But this is your choice. I'm not going to allow a bad attitude to tear this team apart. You've got thirty minutes to go inside, call whoever you need to call, and then come back out. At that point, you either ask me to cut you or you are all in."

Arians then turned away, not knowing whether he had connected with Fournette. For his entire career, dating back to Pop Warner, Fournette had been the focal point of every offense on which he had played. But now he had spent the better part of two months watching the action from the sideline, questioning his future with the Bucs. Arians's voice wasn't tinged with anger or frustration when he spoke to Fournette, but his tone was serious; Arians was fully prepared to release Fournette if the running back didn't drastically change his demeanor and display a more positive, upbeat, engaging attitude. This was Arians's gentle way of shaking the tree to see if a bad apple needed to fall.

Thirty minutes after their talk, Fournette found Arians. "I'm good, Coach," Fournette said. "I'm good. Let's go to work."

"I *know* we're going to need you," Arians said. "Be patient. Keep your head in the game. Your number is going to get called."

Not even Arians could have imagined how right he was.

॥॥॥॥॥॥॥॥

Brady was ruthlessly efficient against the Vikings: he completed 15 of 23 passes for 196 yards and two touchdowns—a 48-yarder to Scotty Miller and a 2-yarder to Rob Gronkowski. He also heaved a Hail Mary into the end zone with one second remaining in the first half. Gronkowski didn't come down with the ball but did draw a 46-yard pass interference penalty, setting up an 18-yard field goal by Ryan Succop. The Bucs built a 23–6 lead by the third quarter. Aided by three missed field goals and an extra point from Minnesota kicker Dan Bailey, Tampa beat the Vikings—a team that had won five of its last six games—26–14. The Bucs were now 8–5 and still the sixth-seed in the NFC playoff picture.

"We control our own destiny," said outside linebacker Shaq

Barrett, who sacked Minnesota quarterback Kirk Cousins twice. "That's all we've wanted, to be able to set ourselves up to be playing football in January, and we're in that position now."

Brady watched the game film twice, critiquing every offensive play, pinpointing what worked, what didn't, and what he needed to do better. Seven days after beating the Vikings, though, the Bucs' offense was as flat as roadkill in the opening half against the Falcons at Mercedes-Benz Stadium in Atlanta, falling behind, 17–0, after two quarters. Brady had been in this position before against the Falcons: Super Bowl LI.

On February 5, 2017, the Patriots trailed the Falcons, 28–3, early in the third quarter at NRG Stadium in Houston, Texas. The lopsided score caused President Donald Trump, a staunch Patriots fan, to turn off the television in the White House. Up in his private box, New England owner Robert Kraft asked his son, Jonathan, "You think Tommy has given up?"

"No fucking way," Jonathan replied.

Brady then led two long drives, one resulting in a touchdown and the other in a field goal. With 9:44 to play in the fourth quarter, Atlanta still led, 28–12. But then Falcons quarterback Matt Ryan fumbled, which set up a Brady touchdown pass to Danny Amendola. After a punt, Brady guided the Pats on a ten-play, 91-yard touchdown drive to force overtime. New England won the coin toss, and Brady did it again: he meticulously dissected the Falcons defense and calmly orchestrated an eight-play, 75-yard drive. Running back James White scored the Super Bowl–winning touchdown on a 2-yard rush.

It was the greatest comeback in Super Bowl history, and as soon as it was over Brady bent down to the turf, where teammates and photographers immediately engulfed him. Brady couldn't move, didn't want to move, he was so overwhelmed with emotion. Even

Bill Belichick joined in a group hug—perhaps the most vivid display of public affection Belichick had ever shown his quarterback.

Brady hadn't forgotten about that epic game against the Falcons—and neither had any of his new Buc teammates.

In the visitor's locker room trailing Atlanta, 17–0, at halftime on December 20, defensive lineman Jason Pierre-Paul screamed at his teammates, imploring them to play smarter, faster, more physically, and with more passion. Pierre-Paul was so animated that a few players couldn't even understand what he was saying. "Jason was heated, man," said one player. "He was trying to get us fired up. He was *really* fired up."

Brady sat in front of his locker and listened, gritting his teeth, staring at the play call sheet. On the sideline before kickoff, Brady gathered the entire offense for one of the few times all season. "When we get the ball, we're going to score every time we touch it," Brady said, his voice rising. "We get the ball, we score—simple as that. Defense will get us stops, and we're not losing this fucking game. You guys hear me!" The players hollered in approval, ready to do whatever it took to make Brady's words come true.

The Bucs had five possessions in the second half—and scored on all five. Four touchdowns and a field goal propelled them to 31 second-half points, and a 31–27 victory. For the final thirty minutes, Brady operated at the height of his powers, displaying surgical precision on throws over the middle, to the sidelines, over the top of the defense, and through the smallest of openings to his receivers. The Bucs took the lead with 6:19 left in the fourth when Brady hit Antonio Brown on a 46-yard touchdown pass—Brown's first score as a Buc. Brady finished 31 of 50 for 390 yards and two touchdowns as the Bucs played their finest half of football of the season. The offense had become what

Arians and Brady had envisioned during the bye: a perfect blend of the Arians Way and the Brady Way. In the second half alone, Brady was 21 of 29 for 320 yards—the fourth highest yardage total in a half in Bucs history.

In the locker room, the players surrounded Brady, more in awe than ever. For every member of the team, something felt different now, like a raging river had been crossed and it had been Brady who guided them to safety on the opposite bank. He had truly become a Buc on this afternoon in Atlanta, the leader of the team who had told his players what they were going to do and how they were going to do it.

"Those thirty minutes are going to be the turning point of our season," Arians told his team after the game. "We've gotten some things figured out on offense and our defense did an out-standing job of giving us a chance. If we play like we did in the second half, there ain't no damn reason we can't win this whole fucking thing. Let's learn from the first half. We need to play an entire game like the way we did in the second half. But damn it, we can do this."

The Bucs had clinched their first winning season since 2016 and were closing in on their first playoff appearance since 2007. They were now 9–5, in second place in the NFC South behind the Saints, and the sixth seed in the playoffs with two games remaining in the regular season. On the quick flight back to Tampa, Brady buried his head in his iPad, watching game film. The playoffs were approaching, and Brady was still trying to figure out every aspect of the Arians offense. Even in the wake of the biggest win of the season, Brady felt a sense of urgency to keep learning, keep growing, and keep grinding.

To Brady, the opening play script read like a masterpiece, a mixture of plays that featured the throws he liked (quick strikes over the middle to receivers on crossing routes and flares to running backs) with the down-the-field passes favored by Arians and Leftwich. The marriage of the Arians and Brady playbook would be strengthened on December 26 at Ford Field in Detroit against the Lions.

After receiving the opening kickoff, Brady eviscerated the Detroit defense, looking as comfortable running the offense as he had all season. First he found Chris Godwin over the middle for 8 yards. Then he lofted a deep ball down the left sideline to Mike Evans for 33 yards. Channeling his inner-Arians, he kept the offense in fifth gear. His third completion of the drive was another 33-yard strike, this one to Gronkowski for a touchdown.

The Lions didn't have their head coach (Darren Bevell missed the game due to COVID-19 protocol), didn't have their starting quarterback for most of the game (Matthew Stafford went out with a lower leg injury on Detroit's first offensive series), and didn't have a chance against the Bucs. On Tampa's second series, Brady continued to stay aggressive, first hitting Godwin deep down the middle for 47 yards and then arching a rainbow to Mike Evans for a 27-yard touchdown. For a team that had been outscored 59–7 in the first quarter in their previous six games, the start to the Lions game went exactly as it was designed by Brady and Leftwich, and was a continuation of the high level of offensive production that the Bucs displayed in the second half against the Falcons six days earlier.

At halftime the scoreboard read: Bucs 34, Lions 0. Brady had been nearly flawless in the first thirty minutes of action, completing 22 of 27 for 348 yards and four touchdowns. Arians pulled his starting quarterback at the start of the third quarter, but later realized that if he had kept him in, Brady could have had a chance to break one of the oldest and most hallowed records in football: Norm Van Brocklin's single-game passing yardage mark of 554 yards, set in the year 1951.

"Tom would have broken that record if we didn't pull him," Arians said. "As a coach, you're usually not thinking of records during a game. But that would have been a cool one for Tom to get. He had a hot hand. It's like a golfer in the zone—the hot quarterback can make every throw in the book from every arm angle."

Tampa beat the Lions, 47–7, to raise their record to 10–5, clinching their first playoff appearance in thirteen years as the sixth seed in the NFC. There was a celebration in the Bucs locker room, but Brady was hardly the life of the party. His eyes were already focused on future games. "There will be a bunch of teams that make it to the playoffs this year and there's only going to be one team that ends up happy," he said. "We're not done."

"[When I got to Tampa] we needed a defense if we were going to win," Arians said. "I knew we'd score points. We've always scored points. It's just a matter of getting a defense. We wanted to keep all those defensive guys here. Then we just added pieces offensively. Tom was huge. Rob was huge. They were more, 'We know how to win.' That's what they brought. I like where we're at now. Anything can happen. I've been a six seed [with the Steelers] and won the Super Bowl. Anything can happen."

History was made in the regular season finale, played at Raymond James Stadium on January 3 against the Falcons: After catching a 20-yard pass late in the first quarter, Mike Evans became the first player in NFL history to begin his career with seven straight seasons of at least 1,000 receiving yards. But just one play later Evans was hurt when he slipped in the end zone.

On the sideline, the coaches fell silent—a rarity for the staff—as they watched the doctors attend to their star receiver; a few players uttered prayers. Evans had become Brady's favorite target and, at six-foot-five, 230 pounds with 4.5 40 speed, he drew constant double teams, which opened up the field for the likes of Godwin, Gronkowski, and Brate. Everyone on the sideline knew that Evans would be vital to the team's championship aspirations.

Evans was placed on a cart and driven to the locker room. Arians received the medical report later and it reinforced to him that this indeed could be a special season: Evans had no structural damage to his knee and would be available for the first round of the playoffs. "You need some luck to win it all," Arians said. "And we got lucky with Mike."

Even without Evans, the Bucs offense hummed against Atlanta. Brady threw often to Antonio Brown, his former houseguest— Brown had moved into a new place with his family—who caught 11 passes for 138 yards and two touchdowns in Tampa's 44–27 victory. Arians and Licht had been roundly criticized for signing Brown because of his previous off-the-field conduct, but once he joined the organization and was under the watchful eyes of Arians and Brady, he was a model teammate. "Couldn't have asked for anything more out of Antonio," Arians said. "He knew that if he made one mistake, no matter how little it was, I was going to cut him. I made that clear. And he responded by showing up, working hard, being a great teammate, and keeping his personal life in order. He knew his role. He knew he wasn't going to be a starter. But he embraced it, really grew, and I think it's a good life lesson here. That's why I've always believed in second chances."

Second chances. For Brady, the season in Tampa was a different kind of a second chance—one to prove he could play again at a high level with another franchise in another offense with another coach. He finished the season with 40 touchdown passes, a Bucs franchise record and the second highest total of his career (50 in 2007). He threw for 4,663 yards (the second most in franchise history) and set the Bucs mark for highest passer rating in a season (102.2).

"When I first talked to Tom after we signed him, I told him that I thought he'd throw for forty touchdowns," Arians said. "And he did just that."

After the bye week and the long heart-to-heart between Arians and Brady, the Buccaneers offense had been on a roll, averaging thirty-seven points during their four-game winning streak. Brady was peaking. The offense was jelling. And the defense was making key plays at big moments. The ingredients of a team capable of making a deep run in the playoffs had come together.

Hours after the game, Arians walked out of the stadium. The Florida sun was setting, streaking the sky in red-orange hues. But his night was just beginning: he was already plotting what the Bucs had to do to win the franchise's first playoff game since 2002—the season, the only season, Tampa Bay hoisted the Lombardi Trophy.

|||||||||||||||||||||||||||||||||||||||||||||||||||||||||||||||||||||||||||||||||||||||||||||||||||||||

CHAPTER 12

The Conquering of Quarterbacks

Chase Young wasn't shy about his intention to rough up Tom Brady. The former standout defensive end from Ohio State told reporters at the 2019 scouting combine that there was one quarterback he wanted to sack more than any other in the NFL: Brady. When Young's Washington Football Team beat the Eagles in Week 17 to win the NFC East and set up a first-round matchup with the Buccaneers at FedEx Field on January 9, suddenly sacking Brady was a possibility. As the rookie pass rusher walked off the field, he yelled, "Tom Brady! I'm coming! I want Tom!"

Brady didn't engage in the war of words; he rarely does.

He had been sacked only twenty-two times in sixteen games, but now he was about to face one of the top defensive lines in the NFL, led by Young, the defensive rookie of the year. The Bucs had planned to lean on its rushing attack against Washington, but in pregame warm-ups starter Ronald Jones tweaked his quadriceps and was scratched from the lineup. This meant that the Bucs' starting running back would be Leonard Fournette.

"I knew we were going to need Leonard at some point," Arians said. "After we had our talk, his approach changed. He still wasn't getting the carries he wanted, but he was a team-first player. This was his chance."

Before the game, Fournette told rookie running back Ke'Shawn Vaughn, "Listen, it's just the two of us. You're going to have to take a couple loads. I'm going to take a couple loads." Fournette did more than that. He rushed for 93 yards and added another 39 receiving. He played a vital role—running, catching, or making a key block in pass protection—in every scoring drive except a first-quarter possession that ended with a 36-yard touchdown pass to Antonio Brown. In the second quarter Fournette kept the chains moving by making two critical receptions on a drive that eventually led to a touchdown and, later, he had a 22-yard reception that helped set up a Succop field goal. Then, late in the game, he lumbered 17-yards to keep a drive alive and capped that with a 3-yard touchdown burst, propelling the Bucs to a messy, ugly—yet to them, exquisitely beautiful—31–23 win.

"You could win 100–0 and it's going to be the same result in the end," Brady said. "You'd love to play great every game—I think it's good to win and advance."

But doing so had been dicey. Tampa had won by a mere eight points, in part because they'd scored just one touchdown in five trips inside the red zone, kicking field goals instead. That kept the game close in spite of Brady's 381 passing yards, a Bucs postseason record. Rookie tackle Tristan Wirfs held Chase Young to zero sacks and zero hits on Brady. Still, after the game, Young approached Brady, tugged on his jersey, and asked if he could have it. Brady politely declined. But a few days later Brady, forever a

Michigan Man, sent the former Ohio State Buckeye a jersey with two words inscribed on it:

Go Blue!

||||||||||||

They continually were spellbound by his attention to detail. For every meeting all season long—and every meeting in the days leading up to the divisional round matchup with the Saints on January 17 at the Mercedes-Benz Superdome—Brady showed up five minutes early and sat in the front row, his eyes locked onto either the screen in front of him or the coach at the head of the room. He always took copious notes in one of his many notebooks, cutting the image of a graduate student in a small lecture hall feverishly jotting down every salient point made by his professor. He carried his notebooks with him everywhere, from meeting to meeting, room to room. At home before retiring for the night, he'd review everything he had written down, converting the words to mental images of action on the field, committing all of it to memory. At age forty-three, there was still one thing that Tom Brady could do as well as any quarterback in the NFL: prepare.

"His habits are just so good that they rub off on everyone else," cornerback Sean Murphy-Bunting said.

"There is a reason that Tom can manage and comprehend a ton of data," Christensen said. "If he's awake and it's during the season, he's basically always focused on football. I mean hyperfocused. He was especially locked in before the Saints game. He knew exactly what we were up against."

The Saints had been Brady's nemesis in 2020. In the season opener, he threw two interceptions in the Bucs' 34–23 loss and appeared entirely flummoxed at the helm of Arians's offense. When the two teams played again in early November, the Bucs suffered their worst defeat of the season, a 38–3 thrashing at Raymond James Stadium that left Brady looking dazed and confused against the New Orleans defense. Now Brady would have his chance at revenge.

Brady had never lost to a team three times in a single season. "The Saints had our number," Arians said. "We played our worst game of the season against them when they beat us 38–3. We were lost in that game and basically nothing went right. But ever since we made that comeback against Atlanta, we had become a different team. We were confident on offense. Tom had command of everything. And on defense we were making key stops and creating turnovers. Games in the playoffs always come down to a handful of plays, and we knew this was going to be the case against the Saints."

The opening offensive plays for the Bucs didn't go as scripted: Tampa went three-and-out on its first two series, with Brady misfiring on his first third-down pass to Gronkowski and then getting sacked for a ten-yard loss on the Bucs' second third-down attempt. But even with the slow start, Brady maintained his stone-faced composure on the sideline. Sitting alongside Christensen on the bench, Brady flipped through images of the Saints defense on an iPad, calmly reviewing what coverages and pass-rush schemes New Orleans was employing. Twice during the 2020 season the Bucs had overcome seventeen-point deficits with Brady leading the way—a feat that Tampa hadn't accomplished even once in a season since 2011—and with him at quarterback anything felt possible. With Brady, no matter the score, the canvas was always blank, always awaiting his brushstrokes. And with him, the deeply held belief on the sideline was that not even a lethargic start against the Saints would derail their charge to the Super Bowl.

"Tom always gives you hope," Christensen said. "And in football, just like in life, hope is a powerful thing."

But New Orleans jumped on the Bucs at the outset, controlling the pace of the game, possessing the ball for eight of the opening eleven minutes. The Saints' Will Lutz kicked a pair of first-quarter field goals as New Orleans seized a 6–0 lead. Then Arians made a crucial call: on fourth and one from the Bucs' own thirty-four-yard line—and facing the prospect of a third straight three-and-out—he decided to go for it. "We didn't come here to play scared," he said into his headset. "Let's go." Brady pushed the ball ahead two yards on a quarterback sneak. The play awakened the slumbering offense. Twelve plays later, Succop drilled a twenty-six-yard field goal, trimming the Saints' lead to 6–3.

Then the defense, which had emphasized forcing turnovers all week in practice, made a play. Cornerback Sean Murphy-Bunting—who twenty-one months earlier had been picked in the second-round of the NFL draft out of Central Michigan, another building block of Licht's roster reconstruction—jumped an out route by Saints wide receiver Michael Thomas and intercepted a throw from Brees. Murphy-Bunting broke a tackle and blazed down the sideline for thirty-six yards, reaching the Saints three. On the next play, Brady hit Mike Evans for six points. "We knew the more we could take the ball and put it in Tom's hands, take it out of Drew's, the better situation we'd be in," Murphy-Bunting said. "We just had an emphasis on taking the ball away anyway we can."

The plan worked. In the third quarter rookie safety Antoine Winfield Jr.—a second-round pick out of Minnesota in 2020, another Licht building block—forced Saints tight end Jared Cook to fumble, which was recovered by linebacker Devin White—a first-round pick in 2020, yet another Licht building block. White returned the fumble eighteen yards to the New Orleans forty-yard line. Five plays later Brady slung a pass to Fournette for a 6-yard touchdown, tying the game at 20. Then, with 7:18 to play in the fourth quarter, White picked off Brees and raced 28 yards before being tackled at the Saints' 20-yard line, setting up a one-yard touchdown run by Brady and giving the Bucs a 30–20 lead. Forty seconds later, safety Mike Edwards—a third-round pick out of Kentucky in 2019, another Licht building block—intercepted yet another Brees pass, sealing Tampa's win.

"The first thing Jason and I wanted to do when I got here was build our secondary," Arians said. "We experienced the dividends of that against the Saints. Our young guys made play after play after play. And when you give Tom the ball in good field position, he's going to make you pay."

Arians slowly walked across the field after the game, the music "Don't Stop Believing" blaring from the speakers. The coach had a sly, half-smile on his face, like a poker player who had just broken away from the table with his pockets full of chips. Even though New Orleans had beaten Tampa twice during the season, when the stakes were the highest, when Arians went all in at the table, he had emerged as the winner. He never stopped believing in himself or his quarterback.

Long after the game, long after earning his record fourteenth trip to a conference championship game, Brady walked alone across the Superdome field, a brown duffel bag hanging from his right shoulder. The stadium was near empty—most of the workers had left for the day—but Brees lingered with his wife and their four children on the far goal line from Brady. The forty-two-year-old Brees was three months away from announcing his official retirement, and he didn't want this moment to end, to let it go, to walk away from the stadium for the final time.

Brady approached and gave Brees a long embrace. Then he hugged Brees's wife, Brittany, telling her that he admired both of them. Brady suspected this was Brees's final goodbye as an NFL player, and now he wanted to share some time with him, to honor him, to tell him of the respect he'd always had for his work ethic, his drive, and his ability to will his team to victory. Brady could see a little bit of himself in Brees, who had been told since college that at five-foot-eleven he was too short to play in the NFL. But just as Brady had overcome the stigma of his draft pedigree, so, too, had Brees proved that in spite of his stature he was a Hall of Fame quarterback.

"I just think so much of him as a person and a player," Brady said. "I know what it takes to do what he's doing, and he knows what it takes to do what I'm doing. I think there is just a lot of mutual respect."

As Brady spoke to Brees and his wife on the Superdome field, Brees's sons Baylen, Bowen, and Callen—each wearing a replica of their dad's number 9 Saints jersey—threw around a football while his daughter, Rylen, did a few cartwheels. Then Brady grabbed a football and instructed Baylen to run a route to the corner of the end zone. As Baylen ran, Brady lofted a perfect strike for an imaginary touchdown. Brady fist-bumped Callen and, as he left, told him, "Be nice to your sister."

Then the Tampa quarterback was gone, strolling away into the darkened concrete catacombs of the Dome, leaving behind the kids and Brees—leaving behind another quarterback whose career ended on the day he was beaten by Tom Brady.

||||||||||||

Two days before the Bucs traveled to Green Bay to play the Packers in the NFC Championship Game, Brady and Christensen spoke on the practice field at One Buc Place. "How do you feel?" Christensen asked.

"Really good," Brady said. "I've got a bead on these guys. We got a great plan. We should be good."

On the morning of the game, the team buses carrying the Bucs players and staff rolled up to historic Lambeau Field, which opened in 1957 and is the oldest continually operated stadium in the NFL. Many of the younger players on the roster had never seen Lambeau, but Arians and Brady reminded them during the week that it was just another football field. Arians had enjoyed a lot of success at Lambeau: as an assistant and head coach, his record at the stadium was 6–2.

Snow had fallen from the Midwestern sky most of the morning, but it had subsided by the time the players emerged from the locker room for pregame warm-ups. The temperature hovered around 30 degrees—warm white clouds shot out of the players' mouths into the cold air—and now as the Bucs prepared to run onto the field they could barely hear each other because of the 7,772 fans. Most were seated in the lower bowl of the stadium, and nearly all held signs that they banged on the metallic bleachers, making more noise than the Tampa players had heard all season at any stadium.

"It sounded like a real NFL game," Arians said. "I never thought that so few people could make so much damn noise, but they did. And that made it difficult for us to communicate on a few occasions, mostly because we weren't used to dealing with the crowd."

Brady's midweek gut feeling about having a "bead" on the Packers defense was prescient. He completed his first six third-down passes (including throws of 27 yards to Evans, 14 to Godwin, and

15 to Evans for a touchdown) as the Bucs raced to a 14–7 lead early in the second quarter. The key sequence of the game took place with 13 seconds left to play in the first half, with Tampa leading, 14–10, and facing a fourth and four from the Green Bay 45-yard line.

At first Arians called for the punt team to take the field, but then signaled for a timeout to reconsider his decision. If the Bucs went for it and failed, it would give Packers quarterback Aaron Rodgers enough time to either throw a Hail Mary into the Tampa end zone or hit a receiver on a 25- to 30-yard throw to set up a field goal attempt. "I'd say 99 percent of the head coaches in the NFL would punt the ball in this situation," Christensen said. "It's just too risky. You've got the lead on the road in a championship game. You should feel good about that. But Bruce is never going to take his foot off the gas pedal."

On the sideline Arians told his offensive group, "We didn't come here not to take chances. Let's do this."

On fourth and four, Brady took the shotgun snap, quickly diagnosed the defense, and checked down to Fournette, who caught the ball flaring out of the backfield for six yards and a first down. Brady signaled for a time out.

On the sideline, Brady huddled with Arians again. The coach wanted to go for what he calls "the kill shot," and Arians and Leftwich called their favorite play: 80 Go. Brady's primary receiver would be Scotty Miller, whose mission on 80 Go was to run straight down the field as fast as he could.

Brady received the shotgun snap. A linebacker blitzed, but Fournette threw his body in front of him, giving Brady an extra second of time to set up and throw. And so he did. Brady unleashed the ball. It arced through the cold Wisconsin sky, spinning beautifully. Miller had run this route hundreds of times, dating back to the Brady-organized workouts at Berkeley Prep in Tampa in April. Brady had analyzed nearly every deep ball he had thrown Miller's way, calculating why he missed some throws and what clicked when they connected. Countless hours of on-field action and off-field study had gone into the making of this moment.

Packers cornerback Kevin King had peaked into the backfield, allowing Miller—who as a college player at Bowling Green ran a 4.36 40-yard dash time—to blaze past him. Brady hit his wide receiver in stride for a thirty-nine-yard touchdown with one-second left on the clock. As Miller crossed the goal line, Brady raised his hands skyward. On the sideline, Arians thrust a clinched fist forward, like a prize fighter throwing a devastating punch. This touchdown, more than any other that was scored during the season and in the playoffs, reflected Arians's approach to football. Brady had always wanted to play for a head coach who never showed fear, who showed belief in his quarterback by allowing him to pick the calls on the play sheet, who took every reasonable opportunity to crush the spirit of the opponent, fourth down be damned. "It's the faith that Bruce shows in you that makes you never want to let him down or not come through for him when he gives you a chance," Brady said.

While Brady struggled in the second half against the Packers—his stat-line after the break read 7 of 14 for 78 yards, one touchdown and three interceptions—the Tampa defense made enough plays to keep the Packers at bay. It started with safety Jordan Whitehead popping the ball out of the hands of Green Bay running back Aaron Jones, leading to an 8-yard Brady touchdown strike to Cameron Brate and a 28–10 score.

Then, with the Bucs holding a 31–23 lead, the Packers moved the ball to Tampa's eight-yard line. But Rodgers misfired on three straight passes. Needing a touchdown and a two-point conversion to tie the game, Packers coach Matt LaFleur opted to kick a field goal with 2:15 remaining in the game. LaFleur was the anti-Arians, a coach so conservative that even several Packer players could be seen on the sidelines with wide-eyed inquisitive, did-he-really-just-call-that expressions on their faces.

The Bucs never relinquished the ball. On a third down Packers cornerback Kevin King tugged on Chris Godwin's jersey as Brady's pass spiraled in Godwin's direction. A late yellow flag landed on the grass field—it was questionable whether the pass was catchable—but ultimately the penalty was assessed to give the Bucs a first down. With two seconds left, Tampa lined up in Victory Formation. Then it was over: the Bucs would be the first team in NFL history to play in a Super Bowl at their home stadium.

Arians was so caught up in the moment—and so ready for his traditional postgame cocktail with Licht—that he forgot about the trophy presentation. He was midway down the tunnel toward the locker room when a staffer told him he was needed on the field. "Oh shit, that's right," Arians said.

As soon as the final whistle blew, Brady beelined it past his teammates, past the cameras, and to a yellow fence at the edge of the sideline that separated the crowd from the players. "Can I say hi to my son?" he asked a security guard.

Jack, thirteen, came running down the concrete steps of the stadium. Jack had been with his dad in Brady's New York City apartment when he signed his contract, and now they were together again, sharing a tender father-son moment. "Love you, kiddo," Brady said. "How about that? We're going to the Super Bowl."

Not far from Brady and his boy, Aaron Rodgers trudged off the field, his head down—another quarterback vanquished by Brady in the postseason. It was Brady's first playoff victory over Rodgers, making Rodgers victim number twenty-seven, an astonishing count.

"I'm just pretty gutted," Rodgers said. "It's a long season. You put so much into it to get to this point . . . It's going to hurt for a while."

In the locker room, Brady spotted linebacker Lavonte David with tears streaming down his cheeks, overcome with emotion. The quarterback walked up to the longest-tenured Buc. "Why are you crying?" Brady said. "We ain't done nothing yet."

Just then, the tears dried up. When Tom Brady is on your team, there's only one time to truly celebrate.

‖‖

CHAPTER 13

Super Bowl LV

It could be seen for miles and miles, glowing high in the dark Florida sky, the image of the quarterback lording over the Tampa skyline. During Super Bowl week the likeness of Brady— in action, hands on ball, preparing to pass—along with other Buc players, was projected nightly onto the Sykes building in downtown Tampa.

In the days leading up to Super Bowl LV, dozens and dozens of curious fans stopped by Brady's house. They'd hop out of the cars, whip out their cell phones, and begin filming the Jeter Mansion. There were so many onlookers that a police officer eventually was called in to stand guard in front of the black iron gate. In a nearby yard stood a sign that read: MY NEIGHBOR IS THE GOAT. It featured a four-color image of an actual goat wearing a Bucs hat.

As kickoff to the Super Bowl neared, the fans kept coming. Two fans were at his house when Brady pulled his Ford Raptor into the driveway. *It's him!* they yelled, as they quickly pulled out their iPhones to record the sighting. After Brady disappeared behind the gate, the fans exchanged high-fives and my-oh-my looks, chuffed with their Bigfoot moment.

Brady stayed alone at the mansion during the two weeks between the NFC championship game and the Super Bowl. Gisele and their two children left the house so Brady could focus on preparations for his record-setting eleventh Super Bowl appearance. He spent extra hours working out in his gym with his body coach, Alex Guerrero, stretching and fine-tuning his body. He was careful not to expend any more energy than was absolutely necessary. He wanted to build up his stamina for the Super Bowl, which his experience informed him was the longest day of the year, spanning from the extended pregame to the prolonged halftime to meeting media requirements after the game.

The most important present Gisele and his kids gave Brady by leaving him in seclusion was the gift of time. In the days before facing the Chiefs, Brady spent virtually every free minute he had watching film. Now, no longer burdened with daily family responsibilities, Brady could lock himself in his home office or living room or bedroom and study the Chiefs defense on his team-issued iPad with full concentration—in silence, no distractions. His face illuminated by the glow of the screen, Brady dissected play after play of the Chiefs, not just from when Kansas City had beaten the Bucs earlier in the season, but also every other game the Chiefs had played in 2020. Brady looked for tendencies and presnap clues that would indicate what coverage the secondary would slip into once the ball was snapped. He looked for "tells"—subtle motions or sounds—transmitted unknowingly by linebackers that would tip him off about what they were about to do; whether, for example, if a blitz was coming or if they were about to drop into coverage. He searched for areas of the field—based on down and distance and defensive alignment—where the

Chiefs could be exploited, and zeroed in on which defenders he wanted to attack.

Brady can remember a vast amount of information—his mind is as agile, according to Arians, as any quarterback he has coached—and Brady will download new data into his brain at all hours. In high school, he used to bring his teammates to his house in San Mateo to watch film while his mother made them lunch. In 2016, just hours after beating Pittsburgh to advance to the Super Bowl, Brady stayed up until 1:30 a.m. analyzing the depth chart of the Atlanta Falcons, pinpointing the Falcons' strengths and weaknesses. He later acknowledged it was unhealthy to stay up about five hours past his normal bedtime—Brady often does cognitive exercises to destimulate and slow down his racing mind, enhancing his ability to drift off to sleep—but he couldn't help himself. The extra hours of studying paid off when he led the Pats to a Super Bowl victory. Would the extra hours now pay a similar dividend?

||||||||||||||

Four days before facing the Chiefs, after practice on Thursday evening, wide receiver Chris Godwin and his fiancée, Mariah DelPercio, slid into Godwin's car and drove toward the flood-lights that illuminated Raymond James Stadium. Barricades surrounded the stadium, so the couple parked and started walking. They took photos and soaked in the scene. Like every Buc, Godwin had always dreamed of this moment, of this game. After several minutes, they embraced and drove away into the night.

After another practice at One Buc Place guard Ali Marpet and tight end Cameron Brate reflected on how far the team had come since Brady had arrived. The previous February, after Tampa had failed to make the playoffs, Marpet threw a Super Bowl party at his house. He invited several of his teammates and ordered enough fried chicken from Publix grocery store to feed a small army. As Brate watched Kansas City defeat San Francisco, 30–21, at Hard Rock Stadium in Miami Gardens, Florida, from Marpet's living room couch in Super Bowl LIV—Mahomes was named MVP—he was overwhelmed with one thought: *We are so far away from playing in this game. I can't even envision it.* But now here they were, only hours away from squaring off against Mahomes and the Chiefs, the same quarterback and team they had seen celebrating on the television screen twelve months earlier. "It's all because of Tom," Brate said. "He's the reason we made it to the Super Bowl. He changed everything."

"Tom is playing for his teammates right now," Arians said before the game. "He wants those guys to experience what he's experienced six times. He truly does. He knows how hard they have worked. I think, personally, too, he's making a statement. It wasn't all coach Belichick who won those Super Bowls in New England."

Brady and Leftwich met every afternoon in the days leading up to the Super Bowl. On the Thursday before facing Kansas City, they held their usual meeting to review the game plan and opening play script. The Thursday sit-down typically lasted an hour during the season, but today they met for less than ten minutes. Both felt confident about the plays they had installed for the Chiefs. "I really, really love our plan," Brady said. "We are good."

"I love this, too," Leftwich said. "We're going to have a good night if we execute the way I think we will."

In his second-floor office at One Buc Place, defensive coordinator Todd Bowles analyzed every play the Chiefs offense had run against the Tampa defense in their Week 12 matchup that Kansas City had won, 27–24. In that game, the Bucs defense had been gashed by Mahomes and Company, surrendering 543 total yards, including 238 yards in the first quarter alone. Mahomes wound up throwing for 462 yards. Wide receiver Tyreek Hill caught three touchdown passes and had 269 yards receiving—the most receiving yards the Bucs had ever given up to a player in franchise history. Bowles had his cornerbacks mostly play man-to-man coverage in Week 12, a strategy he wouldn't employ again in the upcoming Super Bowl. "Todd was sick of hearing about how unstoppable the Chiefs were," Arians said. "He worked his tail off to put together a plan to make sure there wouldn't be a repeat of Week 12."

"We were fortunate that we got to play Kansas City during the season," Bowles said. "You can see their speed on offense on television and on film, but then when you're on the field facing them it's an entirely different thing. We didn't adjust well to their speed in the first game. They jumped on us early. But once we settled down, we played a competitive game. One of our main goals heading into the Super Bowl was to not let them start fast—no quick strikes, no deep completed balls. We wanted to make them earn everything. We thought if we could just make it through the first quarter, we'd be all right."

Unlike in their previous matchup, the Chiefs were now a wounded offensive team. Kansas City's starting left tackle, Eric Fisher, had torn his Achilles in the AFC championship game. The Chiefs' starting right tackle, Mitchell Schwartz, would miss the Super Bowl with a back injury. Bowles believed the Bucs could pressure Mahomes simply by rushing four defensive linemen, even though this went against his career-long tendency to blitz linebackers and safeties. During the regular season the Bucs had the NFL's fifth highest blitz rate at 39 percent. But now with the Chiefs starting two backup tackles, Bowles opted for a more conservative game plan. He would use a two-safety deep approach, which he believed would minimize Mahomes's ability to throw the ball deep.

Bowles would employ a combination of Cover 2 (zone coverage with two deep safeties) and 2-man (man coverage with two deep safeties). Instead of blitzing, Bowles designed a scheme where his linebackers would drop into coverage and the Bucs would shadow Hill with an extra defender. Bowles and Arians weren't worried about Mahomes beating them on scrambles because Mahomes, too, was wounded, having suffered a toe injury in the Chiefs' opening playoff game against the Browns.

"Patrick isn't going to beat us running," Arians told Bowles. "We can let him run all day. We'll just keep chasing him around and see if we can make some plays."

‖‖‖‖‖‖‖‖‖‖‖

It was a regular week of practice leading up to the Super Bowl: the players and coaches stayed at their homes and slept in their beds; the meeting schedule essentially remained unchanged; and the media obligations—aside from a few extra Zoom sessions with reporters—weren't as overwhelming as a typical Super Bowl week. "The familiarity of staying in our normal routine was a big advantage," said Harold Goodwin, Tampa's assistant head coach and run game coordinator. "There was a high comfort level for us."

On the eve of the game, the team moved into the Grand Hyatt Hotel near the Tampa Bay airport. Around 6:00 p.m. that night, in one of the hotel's sprawling ballrooms, the players and coaches feasted on a spread of food befitting a medieval king: a buffet featuring steak, crab cakes, lobster bisque soup, coconut shrimp, and desserts ranging from cake to ice cream to different types of pies. "We had food that I'd never even heard of," Goodwin said.

After dinner, Arians rose from his seat and stood in front of the team. First, he played a highlight video of the season, reminding the players of the journey they had taken to reach this moment. "We worked hard to get here," Arians said at the final team meeting of the postseason. "Enjoy this moment. Conserve your energy. There will be a long pregame tomorrow and a long halftime. We play for each other and we fight like hell for each other. We don't need any heroes. Just do your job. Be smart, fast, and physical. It's hard to make it to a Super Bowl, so try your best to take it all in and play your ass off."

Meeting with his defense later that night, Bowles reviewed their game plan one final time. "Just do your job and don't worry about anything else," Bowles said, practically echoing Arians's words. "No matter what happens, don't look at the scoreboard. Just focus on what is directly in front of you. If something goes wrong on a particular play, forget about it and move on."

On Super Bowl morning, Arians spent quality time with his family. Arians had forty family members and friends in town, including his ninety-five-year-old mother, Kay. The Glazer family had sent a private plane to pick up "Gram" and other family members in Hanover, Pennsylvania, and now she was with her son before the biggest professional moment of his life. "I'm so proud of you," she told Arians. "The Chiefs are going down; the Chiefs are going down."

"Yep Mom," replied Arians, "they're going down, Mom."

Wearing a red Kangol hat, Arians played in the front yard with his three young grandchildren—Asher, Mills, and Brylee. He tossed miniature footballs to them and pushed them in a swing as neighbors out on morning walks waved and wished the coach good luck. Arians then watched the kids glide down an inflatable water slide nestled between two big oak trees in the front yard. For a few fleeting moments, swaddled in the love of his family, Arians didn't even think about the game.

The head of Bucs security, Andres Trescastro, picked up Arians at his house at noon. "Super Bowl Sunday," Arians told Trescastro. "Let's go." The pair then drove to the team hotel, where Arians met his players and his staff. At 4:00 p.m., they boarded three buses and, with the red-and-blue lights of a police escort leading the way, the Bucs were off to play in the Super Bowl.

||||||||||||

At the stadium volunteers stood in concourses holding signs that read: MASK REQUIRED. Notices were posted on fences, street signs, and telephone poles for blocks surrounding Raymond James: *Covid-19 warning. Feeling ill? No entry with a fever over 100.4 degrees.* A crowd of 24,835 would watch the game, including 7,500 vaccinated health care workers who were given free tickets, but this meant there were more cardboard cutouts in the stands (30,000) than there were real people. The Super Bowl would be the 269th NFL game played during the pandemic. During the season, more than 700 players, coaches, and staff members had become infected with COVID-19—only four Buc players would test positive for COVID during the season—but the coronavirus ultimately didn't keep what is the most watched annual event in American sports from being played.

Arians and Leftwich walked out onto the field as pregame warmups commenced. Brady was on the Bucs' bench reviewing plays with a few coaches, asking questions and emphasizing what he thought would work and what he believed wouldn't. He then jogged onto the field to loosen up. As usual, Arians and Leftwich paid close attention to Brady, analyzing his every throw, every movement, making sure he was in perfect rhythm. "He looks ready," Arians told Leftwich. "He looks great. He's as consistent as anybody I've ever been around."

Arians turned his eyes up to the stands and spotted his mom and wife. He was relieved that they had made it to their seats without incident, and he blew them a kiss; they each blew one in return. He then retreated to the locker room to talk to his players one final time before kickoff.

"Be smart, fast, and physical," Arians said, repeating what he had told the players the previous night. "We play for each other and after the first half we'll come back in here and make whatever adjustments we need to make."

The coaches then left the locker room, striding toward the field. The locker room door closed behind them. Then, for the first time all season, Brady rose in front of his locker and began to speak, his voice reverberating across the room. The players fell silent, intently listening to their leader. "When you win this game, you honor your family and you honor your family name," Brady said. "No one can ever take that away from you or your family. Play for your families. Play for their honor. Now LET'S FUCKING GO!"

The players cheered and hollered and yelled; it was the loudest a Tampa locker room had been before a game since Brady had arrived. And now Brady led his believers through the tunnel to the field. Everyone in a Tampa uniform was now ready—ready to win for Brady, for their teammates, for their coaches, and, most important, for their families.

||||||||||||||

The Bucs' first two offensive possessions ended in punts. Trailing 3–0 with a little more than five minutes remaining in the first quarter, Tampa took over at its own 25-yard line, first and ten. Brady then went to work. He threw a 16-yard strike to Antonio Brown. He hit Cameron Brate for 15 yards, then Brown again for five more. Facing second and five from the Chiefs 8-yard line with 37 seconds to play in the first quarter, Gronkowski slipped out on a pass route instead of blocking, which went against the Bucs' tendency when they had been in this down and distance situation throughout the season. The Chiefs defense wasn't expecting Gronk to run down the right side of the field. He was wide-open and Brady lofted a perfectly sublime pass to his long-time friend, who then pranced into the end zone. He windmilled the ball to the turf as Brady ran to embrace him. It was a historic moment: Brady and Gronk had now combined for more postseason touchdowns (13) than any duo in history, surging past Joe Montana and Jerry Rice of the San Francisco 49ers.

With six minutes to play in the second quarter—after the Bucs defense had held the Chiefs offense to a total of 14 yards on the two possessions following Tampa's first score—Brady and Gronkowski connected for another touchdown score, this one a 17-yarder, increasing Tampa's lead to 14–3. Kansas City kicker Harrison Butker then drilled a 34-yard field goal with 1:01 remaining in the first half, trimming Tampa's lead to 8.

"Let's keep going," Arians said into the headset to Leftwich. "Let's try to score here."

Starting at their own twenty-nine-yard line, Brady handed the ball to Fournette, who was stopped for no gain. The Chiefs called a timeout. Brady then rifled a completion to Godwin for eight yards, setting up a third and two. Kansas City coach Andy Reid signaled for another timeout. On third down from Tampa's thirty-seven-yard line, Brady dropped back to pass and calmly found Gronkowski down the right flank for a five-yard gain and a first down. This little five-yard completion, so seemingly insignificant in the grand scheme of the game, would alter the entire dynamic of the Super Bowl.

Brady hustled his team to the line. He took the snap and threw a deep ball to Mike Evans. The pass was incomplete, but cornerback Bashaud Breeland was called for pass interference—a thirty-four-yard penalty that moved the ball to the Chiefs twenty-four-yard line with eighteen seconds left in the first half. Three plays later, with the ball at the one-yard line, Brady spotted Brown in the end zone. The play was designed for Brown to run to the corner of the end zone, but he ran the wrong route: Brown cut to the inside toward the goalpost rather than to the outside to the back pylon. Brady wasn't fazed. All the time they had spent together—living together, practicing together, reviewing film together—had prepared Brady to expect this miscue from Brown. Brady feathered the ball between defenders and hit Brown in the end zone with six seconds to play. For the Chiefs, it was a spirit-sapping sequence of events—one that left them trailing 21–6 at intermission.

"Keep it going," Arians said in the locker room. "The score is 0–0. Keep playing for each other and let's win the damn Super Bowl!"

After a Chiefs field goal to start the second half, the Bucs drove down the field, ending with a twenty-seven-yard touchdown run by Fournette, increasing the lead to 28–9. It wasn't lost on Brady—or anyone in the Bucs organization—that the first four touchdowns in the Super Bowl had been scored by players that the quarterback had personally recruited to play in Tampa. "Just love what they did, what they added to the team," Brady said of Gronkowski, Brown, and Fournette.

Mahomes was in escape mode—usually in a quick backpedal—for the rest of the game. He was sacked three times in the Super Bowl, intercepted twice, and was under constant pressure from the Bucs' four-man front. According to NextGenStats, Mahomes ran a total of 497 yards before throwing the ball or being sacked, zigzagging on the natural grass while trying to avoid Buc defenders. That was the most yards a quarterback had covered behind the line of scrimmage in the NFL all season. Under constant harassment, Mahomes didn't pass or throw for a touchdown for the first time in his eight-game playoff career, was forced into throwing 23 incompletions, failed to lead his team to at least 10 points for the first time in his 54 NFL starts, and finished with a passer rating of 52.3—the lowest of his NFL career.

"They had a good game plan," Mahomes said. "They kind of took away our deep stuff, they took away the sideline, and they did a good job of rallying to the football and making tackles."

"They played a lot of zone, primarily Cover 2, Cover 4," Tyreek Hill said. "We'd rather see man and that's what we game-planned for—a lot of 2-man, a lot of man. Obviously, zone here and there, but Todd Bowles did his thing tonight. He came out and they just had a better game plan."

"Todd had a good plan," Andy Reid said. "He got us. I didn't see it coming, at all. I thought we were going to come in, we were going to play these guys just like we've been playing teams, and it didn't happen that way. . . . I didn't anticipate this happening."

How did the dismantling of Kansas City happen? Bowles anticipated that the Chiefs would only use five-man protection on the majority of their pass plays, which they did on 92 percent of Mahomes's pass attempts (48 of 52). The Tampa front four was so dominating that, even though Bowles called for blitzes on only 9.6 percent of Mahomes's dropbacks, the Bucs set a Super Bowl record for quarterback pressures (29), passing the previous mark set by the Redskins (25) in Super Bowl XXVI on Buffalo quarterback Jim Kelly.

The Bucs also lined up in a two-high safety shell 87 percent of the time (59 of 68 total snaps), which was the highest rate by a Bowles-led defense in any game during the last five seasons. This effectively eliminated the Chiefs' vaunted deep ball passing attack. During practices leading up to the game, Bowles repeatedly instructed his defensive backs not to allow the Chiefs' playmakers to get behind them. He emphasized tackling in open space and constantly had his players shifting in the secondary in the moments before the ball was snapped, disguising coverages and confusing Mahomes.

"The biggest thing was trying to take away [Mahomes's] first read," Bowles said. "Patrick can run and make plays with his feet. But we didn't want him just sitting in the pocket, zinging dimes on us all day, either. The D-line got some pressure on him, was making him uncomfortable, and that was the key for us."

The last hour of the game in real time turned into an extended celebration on the Buc sideline. As the clock wound down, Arians and Brady embraced on the sideline. "I love you, bro!" Brady said. "That was amazing."

"I knew this was going to happen," Arians responded. "This is why we came together."

The final score: Tampa 31, Kansas City 9.

After the final play of the game—a Victory Formation kneel-down by Brady—a few Buc players doused a bucket of blue Gatorade on Arians, who at 68 years and 127 days became the oldest coach to win a Super Bowl. Arians and Brady had achieved the rare and the remarkable: together they had won eight consecutive games, including back-to-back-to-back victories against three of the top quarterbacks in the game—Brees, Rodgers, and Mahomes. Brady, who finished 21 of 29 passing for 201 yards and 3 touchdowns, also made history. Not only did he win his unprecedented seventh Super Bowl, but he also became only the second quarterback, after Peyton Manning, to lead two different teams to the Lombardi Trophy. At age forty-three, he had now participated in 18 percent of the Super Bowls ever played, and he also became the oldest starting quarterback to win a Lombardi Trophy and a Super Bowl MVP Award.

"Not bad for two old guys," Arians said. "The Fountain of Youth really does exist in Florida."

||||||||||||

In his younger days, Brady used to look for his dad after Super Bowl victories. But now, in the frenzied aftermath of their world championship, Brady walked around the field, hugging his teammates, smiling luminously, searching for his wife and kids. His oldest, Jack, was the first to run into Brady's arms. "Dad!" The father held his boy tight, as if he would never let go. Then Gisele and their younger children, Vivian and Ben, joined in the group embrace. Once they had all shared a moment, Gisele asked her husband, "What else do you have to prove?" But Brady didn't answer her, not here, not now, not when he was reveling in the glow of yet another Super Bowl victory.

But it was a legitimate question. Now that he had won his seventh championship, Brady had leapfrogged two of the most prolific winners of all time: Michael Jordan and Bear Bryant, who each had won six titles in their careers. Brady had cemented his place among the great athletes in history, alongside the likes of Jordan, Muhammad Ali, Joe Montana, Wayne Gretzky, Pele, Babe Ruth, and Jim Thorpe.

What else do you have to prove? For Brady the answer is nothing and everything. When Brady was in his prime, the game appeared to come so easy to him, like he was in a race driving a Ferrari and everyone else in the league was in a Pinto. Couple that with his J.Crew looks and his supermodel wife, Brady became a figure that was larger-than-life, an NFL quarterback with whom so many fans felt they had nothing in common. But this season had been different. This had been a struggle. There had been moments, so rare in his career, when Brady had seemed almost frail. And many moments when he seemed more relatable, more real, more authentic than ever before. More vulnerable. Suddenly, fans could see how he was fighting like hell to hold on to his youth and his boyhood dreams, and to enjoy at least one more season in the sun.

What else do you have to prove? "I want number eight," he said. "Gotta keep pushing."

|||||||||||||

They stood inside Arians's office next to the locker room, the coach and the general manager. Many of the players had already headed to the post–Super Bowl party at the Florida Aquarium, which the team had rented out. But now Arians and Licht lingered together, smiling for pictures and hugging like brothers.

"Not bad for two old bartenders," Licht said. "We've both come a long way."

"Damn right we have, brother," Arians replied. "And we're not done yet."

The two then walked outside into the dark Tampa night—side by side, the coach and the team architect—and then climbed onto a bus that would take them to the party.

CHAPTER 14

||||||||||||||||||||||||||||||||||||||||||||||||||||||||||||||||||||||||||||||||||||||||||||||||||||||||

The GOAT Boat

It was a postcard-perfect afternoon in Tampa, a Chamber of Commerce kind of day: a sun-soaked 83 degrees, a robin's egg blue sky, and a gentle breeze off the bay that swished the fronds of palms throughout downtown.

Three days after winning Super Bowl LV the Tampa players, coaches, and staff boarded four buses at One Buc Place at 11:30 a.m. Arians had said on the morning after his greatest professional achievement that he was disappointed that the Buccaneers wouldn't be feted to a winner's parade due to the strictures of the pandemic. Arians had always wanted to ride in the back of the truck, swill a few beers as the head coach, and interact with the fans. Arians's post–Super Bowl parade history wasn't stellar. In 2009 a group of intoxicated Steeler fans booed him during their parade following Super Bowl XLIII, upset with his too-aggressive play calling as the team's offensive coordinator. One fan yelled, "Get a fullback!" Arians hollered back: "Never!"

But now, less than seventy-two hours after the Bucs won the Lombardi Trophy, the mayor of Tampa, Jane Castor, reacted to

Arians's wish. She announced plans for a flotilla parade on the Hillsborough River. "We had to do it today with the Buccaneers here," Castor said. "A lot of them are leaving town so we just had to do this as quick as we could. We will have an appropriate parade later down the road when we're out from under this pandemic."

As the buses carried the players and coaches toward Tampa's downtown Riverwalk, a few players noticed a tweet that the Bucs had sent out hours earlier. It was the picture of a puzzled Brady holding up four fingers in the Chicago game, when he was confused about what down it was late in the loss to the Bears on October 8. The caption read "How many playoff games did you win this year, Tom Brady?" The players on the bus could only laugh.

After the buses parked at Riverwalk, the coaches and players—many carrying their own bottles of booze—boarded several different boats. Gronkowski, a veteran of postseason victory parades, downed bottle after bottle of water as he strode toward an assembly of boats of various shapes and styles, hydrating himself before the real partying began.

Brady celebrated in style. He captained his $2 million, custom-built boat named *Viva a Vida*. With his youngest son on his lap, he steered the GOAT Boat—as some called it—through the calm blue waters. Built in the Netherlands, the 53-foot, Wager 55-S yacht has three engines that can generate up to 2,000 horsepower. If Brady mashed the throttle, the GOAT Boat could go as fast as 40 knots—or 46 mph—and it has a range of 400 miles. The image of Brady as the ship's captain revealed how far he had come since his early days in Tampa, when he had unwittingly wandered into a stranger's house and had been kicked out of a public park for violating the state's stay-at-home order. Brady now just wasn't another snowbird who migrated year after year from the Northeast; he was a full-on Floridian. He was home.

Brady thoroughly enjoyed himself along the parade route, waving to the thousands along the Riverwalk and acknowledging the fans on boats, jet skis, kayaks, canoes, paddleboards, and even a floating tiki hut. Arians sat cross-legged on the bow of another boat—a coaching Buddha with a Bud Light—pointing and acknowledging the crush of admirers. "I swear that every boat in Tampa was at that parade," Arians said. "There were boats for as far as you could see. I don't think any of us ever experienced anything like that before. That was when it really hit us that we meant so much to the city."

"It was the best party I've ever been to," Leftwich said.

Brady's final pass of the season was his riskiest. Emboldened by tequila, Brady clutched the Bucs' newly acquired sterling silver Vince Lombardi Trophy—twenty-two inches high and weighing seven pounds. Standing at the stern of his boat, he gathered his balance and swung the trophy underhand back and forth, back and forth, gauging the ten-yard distance of water between his boat and one carrying several wide receivers and tight ends. As Brady prepared his underhanded toss, his eight-year-old daughter Vivi yelled—in the loudest voice of reason on the two boats—"Daddy, nooooo!"

But daddy did. He flung the Super Bowl Trophy in the direction of tight end Cameron Brate, who was shirtless. The trophy arched over the sun-sparkled water that was eighty feet deep, its silvery surface shimmering in flight. Brate reached up and snagged the trophy, catching it clean. "That was the best catch of my life," Brate said. "But if I had dropped that? I think I would've had to retire." Brady later said he wasn't thinking when he made the throw, that the tequila told him it was the thing to do. Brady has been known to enjoy himself at Super Bowl parades—according to a former New England teammate he once bested the entire offensive line in a beer-chugging contest—but he never had *this* much fun before. Neither had Gronkowski, who, on another boat, had also shed his shirt and was dancing the afternoon away with several teammates.

The armada followed a V-shaped route through Tampa's river and entertainment district. After hours on the water, the sun-baked, liquored-up players and coaches disembarked their boats at Port Tampa. Like many after a booze cruise, Brady wobbled off his yacht. Backup quarterback Ryan Griffin held Brady up as the two reached land. The video of an intoxicated Brady went viral on social media, but then a funny thing happened: it actually endeared him even *more* to fans. It showed a humanness about Brady that few had witnessed, a sign that even one of the most disciplined athletes in NFL history can, in fact, be overserved like any of us when we are having a good time. The intensely private quarterback had let down his guard, and most fans—even those who didn't cheer for Brady—liked what they saw.

Brady trended on Twitter. He later retweeted the video and made fun of himself, writing, "Nothing to see her . . . just little avocado tequila" in a purposefully misspelled tweet, mimicking his drunkenness in the video.

Brady had been scheduled to say a few words on a stage after the parade, but he was in no condition to do so. Licht spoke for the team when he proclaimed to the crowd, "We're going to fucking win this thing again!"

Arians, with a beer in one hand, told the fans, "We have the best coaches in the NFL, and we've damn sure got the best players. We're going for two and we ain't stopping. We're going to keep this band together."

Arians was hoping to go on one more tour with his band of coaches and players—one that would again end with them all partying like rock stars.

‖‖‖

CHAPTER 15

The Work
Begins Again

*F*ive weeks later . . .

He gazes out a living room window of his five-bedroom home in South Tampa, the sixty-eight-year-old NFL head coach wondering about the future. He sees towering oak and cypress trees with Spanish moss hanging from the limbs like long graying beards. His yard is dotted with kid-sized footballs that the coach last threw to his grandchildren on the morning of Super Bowl LV—a startling sight to couples out on walks who shot oh-my-God-it's-really-him looks in his direction. In the distance, the last blush of sunlight sweeps across the emerald waters of Hillsborough Bay and, beyond that, the Gulf of Mexico, the western horizon afire and aglow as the Florida sun sinks away. For Bruce Arians, the approaching night is rich with possibility: the winningest quarterback in NFL history has just made a decision.

At 6:47 p.m. on March 12, 2021, a black Mercedes 550 swings into the wide driveway of Arians's home in a quaint, distinguished Tampa neighborhood. Behind the wheel is Jason Licht, the general manager of the Tampa Bay Buccaneers. A little more than a month has passed since the team that Licht meticulously constructed—brick by brick, draft pick by draft pick, free agent by free agent—and Arians coached, won the fifty-fifth Super Bowl at Raymond James Stadium.

The game is still fresh in Arians's mind—and now its outcome is indelibly inscribed on his body. Earlier on this day Arians spent three hours laying facedown on a massage-like table at a tattoo parlor a few miles from his house. A tattoo artist carefully crafted the image of the Vince Lombardi Super Bowl LV trophy onto Arians's left shoulder blade. The red-and-black-and-pewter-colored ink, which includes the game's final score of 31-9, is his only tattoo. For Arians, the dream of this body art has been stirring for over four decades, first seeded in 1989 when he began his career in the NFL in 1989 with the Kansas Chiefs as the running backs coach. "Didn't even hurt," Arians says. "I'm still too fucking happy to feel the pain."

Now Licht is on the phone as he parks his Mercedes in the driveway. A few minutes pass and then, still talking into his cell, Licht flashes the thumbs-up sign at Arians. The news is good: he has finalized a contract extension with Brady, one that will keep him in Tampa at least through the 2022 season.

Once off the phone, Licht strides into the house—there is no need to knock; Arians has an open-door policy with everyone in the Tampa organization—and he then hugs his coach. "We got him and we got all the numbers figured out," Licht said. "Tom is so pumped. He's all in on winning it all next season."

Moments later, Arians receives a text from Brady: *Let's fucking go! We're only getting started. Time to get to work!*

The coach and general manager high-five again, smiling like

two kids on the sandlot who just hooked up to complete the most important touchdown pass of their lives. What a moment this is for Arians, Licht, and the entire Tampa organization. The player who just twelve months earlier had been the most coveted free agent in NFL history is now likely to retire as a Buccaneer; the deal keeps Brady under contract through his forty-fifth birthday, the age that Brady wants to reach before he walks away from the game with more Super Bowl rings (currently seven and counting) than any player to ever play the sport. And by restructuring his contract and adding additional years to it—on paper the contract is a four-year extension but in reality it's for two years—Brady's new deal has saved the team $19 million in salary cap space.

In the coming weeks Licht will manage to retain all twenty-two starters from the Super Bowl—the first time a Lombardi Trophy winner returns every starter since the 1977 Oakland Raiders. The list of players that will return to Tampa Bay includes free agent linebacker Lavonte David, free agent wide receiver Chris Godwin, free agent linebacker Shaq Barrett, free agent defensive tackle Ndamukong Suh, free agent running back Leonard Fournette, and free agent wide receiver Antonio Brown. No team has won back-to-back Super Bowls since 2004 when Brady led the Patriots to their second consecutive title, but the reigning champions—the team that won its last eight games of the 2020 season—may still be in the conquering days of its empire.

"I guess this means I'll be coaching in Tampa for at least two more years," Arians says, looking at his wife, Chris, who at one time had to be convinced by her husband that it was a good idea for him to return to coaching in the NFL after spending a year in retirement. "It's going to be a hell of ride."

Chris shakes her head, a soft grin lighting her face. "Well, we can't leave now!" she says. "Tom has been so good to us. He's so genuine. We can't miss out on two more years with him. It's party time!"

Arians and Licht walk over to the kitchen bar, each pouring a victory drink for himself; Arians is a Crown-and-Coke man, Licht opts for vodka and lemonade. As the coach and general manager sit at the kitchen table and chomp on pepperoni pizza from a joint called Marcos, they both reflect on the events of the previous year—from the luring and landing of Brady to the mid-season struggles to the run to the Super Bowl. The stories flow as fast as the cocktails deep into this cool spring night.

For hours here at the kitchen table, Arians and Licht repeatedly close their eyes. On the grainy film of memory, they see it all again: a replay of the season that changed their lives—the season they spent in the sun.

||||||||||||

There he was, striding through a driving rainstorm toward an event hall at Innisbrook Resort in Palm Harbor, Florida, on April 18. Cameras flashed like lightning as he emerged from the cool evening darkness and, cell phone in hand, walked toward the glass doors. Inside, Arians—outfitted in a Kangol hat, a white-and-tan suit, and blue tennis shoes—was for waiting for him. The sharply dressed head coach acted as the quarterback's personal doorman, welcoming him and giving him a hug. "The man has arrived," Arians said as a few women pushed each other to get closer to the duo, hoping to snap a picture with their phones.

In the previous weeks Brady and his family had lit a blue streak across the globe, spending time in Costa Rica and Qatar, where they visited a museum and gazed at a sculpture named "7." But now Brady was back with his coach, and after a brief talk with

Arians in a private room at Inverness Hall and Ballroom, Brady then signed autographs, smiled for pictures, and glanced at the Super Bowl trophy, which sat under a glass case at the end of a long hallway. He then took a seat at a ballroom table for an Arians Family Foundation benefit dinner and auction. Nearby was Rob Gronkowski, who entertained guests at his table with tales from Tampa Bay's Super Bowl season. A few feet away Jason Licht took in the scene, reflecting on how the signing of Brady had transformed a team and a town.

"He is everything and more than we thought he was," Licht said. "It took an entire organization to win the Super Bowl, from ownership to coaches to players to staff members to the people who cleaned the locker rooms, but it all started with Tom. He is the one player in the league who, just by his presence, can change an entire culture."

Brady listened as speakers such as Bucs co-owner Darcie Glazer Kassewitz thanked him for coming to Tampa. Brady then rose from his chair, embraced more coaches and players who were in attendance, and began walking back to his black Ford Raptor, parked outside. Before he reached his truck, Leftwich stopped him. They discussed how the offseason had been going, the players the team had signed, and what they needed to do to get ready for the 2021 season. "I'm ready to get after it," Brady said. "I'm ready for us to try to make another run."

Then Brady disappeared into the darkness, climbing into his truck and driving away into the stormy Florida night. Bolts of lightning fractured the sky; thunder boomed. Party time was over. The work—the thing Brady does best, what he's always done best—was about to begin again.

AUTHOR'S NOTE

||||||||||||||||||||||||||||||||||||||||||||||||||||||||||

This book draws upon my own reporting over a number of years, as well as extensive interviews with many of the key figures within the Bucs' organization. In addition, I am indebted to a number of reporters from around the country who wrote detailed stories about Tampa Bay's march to Super Bowl LV. I want to acknowledge six writers whose works I especially leaned on in crafting this book: Seth Wickersham, Greg Auman, Jenna Laine, Sam Borden, Greg Bishop, and Jenny Vrentras. I also incorporated published material that appeared in my books *The Quarterback Whisperer* and *Chasing the Bear*, and from magazine stories that I wrote for *Sports Illustrated* and *Bleacher Report*. For background on Tom Brady, Rob Gronkowski, and Bill Belichick, I used information from Dan Wetzel's book, *Epic Athletes: Tom Brady* (Henry Holt and Company, New York, 2019), Michael Holley's *Belichick and Brady* (Hachette, New York, 2016), Jeff Benedict's *The Dynasty* (Avid Reader Press, New York, 2020), Charles P. Pierce's *Moving the Chains* (Farrar, Staus and Girous, New York, 2006), Jeff Schober's (along with the Gronkowski family) *Growing Up Gronk* (Houghton Mifflin Harcourt, Boston, 2013), and Howard Stern's interview with Tom Brady on *The Howard Stern Show* on SiriusXM on April 8, 2020.

ACKNOWLEDGMENTS

||||||||||||||||||||||||||||||||||||||||||||||||||||||||||||||||||||

'm so in debt to Bruce Arians. In 2017 we co-authored *The Quarterback Whisperer: How to Build an Elite NFL Quarterback*, and we grew close during the writing process. We played golf together, had dinners together, and spent time with each other's families. Over the years we've maintained our friendship, and this book never would have been possible without Bruce's willingness to open up and share the inside story of the Buccaneers' run to Super Bowl LV.

This project also never would have happened without the leadership of Mike Fetchko, the president and managing director of ISM USA, a sports and marketing consulting firm. "A Season in the Sun" was his idea. A lawyer, Mike and Arians have been friends since they both worked at Temple in the 1980s—Mike resigned from his position as a senior associate athletic director the day Arians was fired in 1988—and he remains one of BA's most trusted confidantes. Mike helped set up interviews, invited me to stay at his Florida home, and frequently called to offer encouragement and insight. From the bottom of my heart, thank you, Fetch.

I sat for hours with Jason Licht, Tampa's general manager.

Jason and I both spent time in our youth living in Lincoln, Nebraska, and we have several mutual friends. When speaking with Jason, it felt like I was talking to someone I had known for years. Put simple, he is Nebraska nice, and was always willing to give me as much time as I needed.

I talked to as many of the Tampa coaches as I could during the reporting process, including offensive coordinator Byron Leftwich, defensive coordinator Todd Bowles, quarterbacks coach Clyde Christensen, assistant head coach Harold Goodwin, tight ends coach Rick Christophel, speed and conditioning coach Roger Kingdom, and several others. Each one was beyond generous with their time. I also spoke to over a dozen players—some on the record, some on background—and all of them patiently answered my questions and opened up to me about their experiences from the 2020 season.

My editor at HarperCollins, Mauro DiPreta, captured the vision of this book based on a five-page memo I sent him a few days before Super Bowl LV. This project required a very quick turnaround, and Mauro's unwavering support and guidance were instrumental. Nick Amphlett, an editor at HarperCollins, performed a pitch-perfect line edit and his big-picture ideas were always on the money. And my literary agent, Richard Pine, also championed this project and helped get the wheels moving.

My stepfather, Gordon Bratz, edited a draft of the manuscript and his careful read improved the narrative one paragraph at a time. A retired Army colonel, Gordy's literary fastball is as lively as ever. I'm so grateful Gordy is in my life. Ira Kaufman, a long-time sportswriter in the Tampa area, lent a graceful editorial hand in shaping elements of the narrative and Joel Poiley, a Tampa-based reporter, pitched in with extensive research, editing, and fact-checking help. Ira and Joel both were instrumental players in transforming this project from an idea into a book. Cole Thompson, one of my former students at the University of Alabama, also helped dig into the backgrounds of several key characters in the book.

Finally, a note to my three small children: Lincoln, 6, Autumn, 4, and Farrah, 4. Writing can be a very solitary process, and it often requires time away from your family. I couldn't be prouder of my three little ones. Even at their young ages, they each are displaying the character and kindness and joy for life that would make any parent proud. My most important title will always be Daddy, and now I've got a lot of lost reading time to catch up on with my beautiful kiddos.

INDEX

||||||||||||||||||||||